International Linear Collider (ILC)

The next mega-scale particle collider

International Linear Collider (ILC)

The next mega-scale particle collider

Alexey Drutskoy

P.N. Lebedev Physical Institute of the Russian Academy of Sciences, Moscow Institute of Physics and Technology, Moscow Engineering Physics Institute, Moscow, Russian Federation

Morgan & Claypool Publishers

Rights & Permissions
To obtain permission to re-use copyrighted material from Morgan & Claypool Publishers, please contact info@morganclaypool.com.

ISBN   978-1-64327-326-6 (ebook)
ISBN   978-1-64327-323-5 (print)
ISBN   978-1-64327-324-2 (mobi)

DOI   10.1088/2053-2571/aae221

Version: 20181101

IOP Concise Physics
ISSN 2053-2571 (online)
ISSN 2054-7307 (print)

A Morgan & Claypool publication as part of IOP Concise Physics
Published by Morgan & Claypool Publishers, 1210 Fifth Avenue, Suite 250, San Rafael, CA, 94901, USA

IOP Publishing, Temple Circus, Temple Way, Bristol BS1 6HG, UK

I dedicate this book to the people most dear to me: my wife and my child.

Contents

Preface

The International Linear Collider (ILC) is a mega-scale, technically complex project, requiring large financial resources and the cooperation of thousands of scientists and engineers from all over the world. Such a big and expensive project has to be discussed publicly, and the planned goals have to be clearly formulated. The demand for the project, motivated by the current situation in particle physics, will be advocated in the book.

The natural and most powerful way of obtaining new knowledge in particle physics is to build a new collider with higher energy. Under this approach, the large hadron collider (LHC) was created and is now operating at a world record center-of-mass energy of 13 TeV. Although the design of colliders with energy of 50–100 TeV is discussed, the practical realization of such a project is not possible sooner than in 20–30 years. Of course, many new results are expected from the LHC over the next decade. However, we must also think about other opportunities, in particular about construction of more dedicated experiments. There are many potentially promising projects, however, the most obvious possibility to achieve significant progress in particle physics in the near future is the construction of a linear e^+e^- collider with energies in the range of 250–1000 GeV. It has been proposed that such a project, the ILC, is to be built in Kitakami city, in Japan. Why this project is important and which new discoveries can be expected with this collider will be discussed in this book.

In the first chapter of this book, the current situation in high-energy physics and the potential experimental approaches in the search for phenomena beyond the Standard Model will be discussed. The physics potential of the ILC will be compared with that of other experimental projects, and the advantages of the ILC will be explained. In the second chapter we discuss the design of the ILC, and the detectors ILD and SiD proposed for the ILC. Also, the alternative e^+e^- colliders will be briefly discussed: the linear collider CLIC with its energy up to 3 TeV, being developed at CERN; the circular e^+e^- collider CEPC with an energy of 240–250 GeV, proposed in China; and a new circular e^+e^- collider FCC-ee with energies 90–400 GeV, proposed at CERN. In the third chapter, the physics program for the ILC will be discussed in detail, with some focus on potential studies, which can be performed at a center-of-mass energy of 250 GeV, planned at the first stage of the ILC project. Special attention will be paid to the studies of the Higgs boson, whose properties can be measured at the linear collider with a high accuracy. The possible measurements of the properties of the Higgs boson at total energies of 250 GeV and 500 GeV will be compared. Later in the third chapter, we will discuss other potential areas of research that can be explored with the linear collider. These topics include a direct search for particles of New Physics, a detailed study of the properties of the top quark, and a precision study of the properties of gauge bosons and their interactions. In conclusion, the most important questions that can be answered with the ILC will be briefly summarized. Also, the current status of the ILC and the prospects for its construction will be discussed.

Acknowledgements

I wish to thank my colleagues for their help during my work on this book. I am grateful to Daniel Jeans for his comments on the text. Also, I wish to thank the staff at IOP Publishing and at Morgan & Claypool Publishers for their careful and detailed editing of this book.

Author biography

Alexey Drutskoy

Alexey Drutskoy started working for the Moscow Institute for Theoretical and Experimental Physics (ITEP) in 1981 after graduating from the the Moscow Engineering Physics Institute. He joined the ARGUS collaboration, DESY (Hamburg, Germany) in 1989 and completed his PhD based on obtained there results in 1993. Between 1993–2001, he worked for the H1 collaboration (DESY), and later joined the Belle collaboration (KEK, Tsukuba, Japan, 1999–2015). He is now a member of the D0 collaboration (Tevatron, Batavia, US), and the ILD collaboration of the International Linear Collider (ILC), which is proposed in Japan. He worked as an international visiting scientist in DESY, Germany between 1995–2001, in KEK, Japan, in 2001, in the National Taiwan University in 2003, and in the University of Cincinnati, USA, between 2003–2010. After a few years of working on the Belle collaboration, he proposed to take data at the $\Upsilon(5S)$ resonance energy to study B_s mesons. His proposal was successfully approved and he played a leading role in the realization of that project. The project was a major success, yielding many new results, which were published by the Belle collaboration using the $\Upsilon(5S)$ data. In 2011 he returned to Moscow and defended his second (doctorate) dissertation on the $\Upsilon(5S)$ and B_s meson studies at B factories. He is now the lead scientist at the Lebedev Physical Institute (LPI) and a Professor in the Moscow Engineering Physics Institute and the Moscow Institute of Physics and Technology.

International Linear Collider (ILC)
The next mega-scale particle collider
Alexey Drutskoy

Chapter 1

Introduction

The International Linear Collider (ILC) (figure 1.1) is a mega-scale project in the field of particle physics, which is planned in Japan and has been widely discussed over the past 15 years. Recently, it was proposed to build the collider in the vicinity of the city of Kitakami in the north of the main island Honshu. This linear e^+e^- collider is envisaged to become the next largest project after the Large Hadron Collider (LHC) in the field of high-energy physics. The final decision on the construction of the project should be made soon by the government of Japan. The demand for such a collider is definitely motivated by the situation that we have today in the field of particle physics.

Figure 1.1. Art image of the ILC in Japan. © Rey.Hori/KEK

doi:10.1088/2053-2571/aae221ch1

1.1 Completion of the Standard Model

During recent decades, the Standard Model (SM) has been developed theoretically and tested experimentally with high accuracy. Today, the SM describes the world of elementary particles and their interactions amazingly well. Figure 1.2 shows the 'zoo' of the SM elementary particles, which includes three generations of quarks and leptons and a set of four particles, which provide interactions caused by three fundamental forces. This set contains a massless photon responsible for the electromagnetic interactions; heavy Z and W bosons responsible for the weak interactions; and a massless gluon responsible for the strong interactions. These four particles are used as carriers for the corresponding fundamental forces. Apart from it in the SM stays the Higgs boson, which provides a mechanism to generate the masses of elementary particles. Without the Higgs boson, elementary particles in the SM would be massless. All elementary particles shown in figure 1.2 have mirror antiparticles with exactly the same properties, and an opposite sign of the charge; for example, the electron e^- is the antiparticle of the positron e^+. In the case of a zero charge, the particles and their antiparticles are indistinguishable. It has to be mentioned that the quarks and leptons have a spin $J = 1/2$; the photon, gluon, and gauge Z and W bosons have the spin $J = 1$; and the Higgs boson has the spin

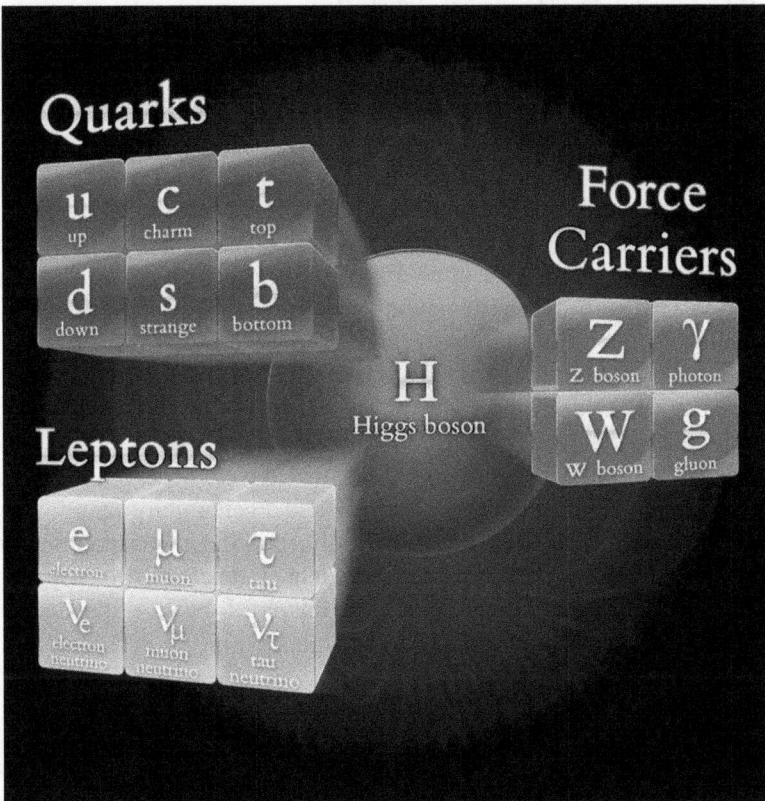

Figure 1.2. The elementary particles of the Standard Model. Source: Particle Data Group.

$J = 0$. The Standard Model is essentially based on symmetry principles. All diversity of the microworld phenomena is represented by this short list of elementary particles, which interact by means of the three fundamental forces: electromagnetic, weak and strong. The gravitation force has a tiny coupling constant, comparing with other forces, and is not included in the SM.

When we say an 'elementary' particle, we assume that the object has no spatial size or internal structure. In fact, this means that, with the current accuracy of experimental instruments, we cannot see the structure of the object, or, in other words, we cannot observe experimentally its components. The atom was assumed to be an 'elementary' object until the moment when its nucleus was observed by Rutherford in an experiment with alpha particles scattering. Later, it was found that the nucleus consists of protons and neutrons, which for some time became 'elementary'. When the structure of protons and neutrons, as well as other hadrons, was observed, we began to call quarks and gluons 'elementary'. Whether the particles shown in figure 1.2 are really elementary or composite, we do not know. From one side, quarks and leptons line up in three generations with repeating behaviors, which may provide a hint toward an intrinsic structure. From another side, a huge difference in the masses of the three generations and a very low experimental limit on the spacial size of the quarks and leptons of about 10^{-18} cm contradict the possibility of having an internal structure within quantum mechanics theory. This indicates the probability that quarks and leptons, as well as photons, gluons and bosons of the SM, are really 'elementary' particles. It should be noted that even elementary particles can have some kind of 'virtual' structure in quantum mechanics.

By today, the SM has been completed theoretically and confirmed experimentally with high accuracy. The last particle, predicted in the framework of the SM, was the Higgs boson, discovered at the LHC in 2012. This was an impressive triumph for the SM. However, there are a number of problems that cannot be solved within the SM. First of all, the SM cannot describe some astrophysical observations, such as dark matter and the asymmetry of matter–antimatter in our universe. There is no answer to the question of why the mass of Higgs boson is not huge, since divergencies in Higgs mass calculations appear within the framework of the SM. It is important to find an adequate theoretical explanation to address this question. Gravitation is not included in the SM at all. The neutrino masses, which were observed experimentally, are difficult to explain within the SM. There is also a general argument: the SM has many free parameters measured experimentally, but these parameters cannot be obtained within the SM from basic principles like symmetries, time–space properties, causality, or any others. Therefore, a more universal theory of particle physics is required, which will include the SM as a part.

1.2 New Physics: phenomena beyond the Standard Model

There are many theoretical models beyond the Standard Model (BSM). Particles and processes, which do not exist in the SM but appear in BSM theories, are often called New Physics (NP). The supersymmetric (SUSY) extension of the SM was proposed a long time ago, and the SUSY was a dominant BSM theory for many years. An important advantage of this theory is its elegant way of cancelling divergences in the mass of the Higgs boson, adding heavy supersymmetric partners to all SM particles, with similar behaviors and different spins. The divergencies caused by the SM and SUSY particles will have opposite signs and will mostly cancel in the calculations of the Higgs mass, resulting in a low Higgs mass of about a hundred GeV, which corresponds to the weak scale. It was predicted that the new SUSY particles will have masses in the range from a few hundred GeV to a few TeV. Moreover, the SUSY particles will be good dark matter candidates, assuming that the lightest SUSY particle mass will be around several hundred GeV or less, and interactions between the SM and SUSY particles are very weak. Some SUSY models offer a weak-scale baryogenesis, and sources of large CP-violation, which are needed to explain matter–antimatter asymmetry. There are SUSY models that include gravitation. Therefore, SUSY is a very attractive theory, which allows one to solve almost all known problems related to phenomena beyond SM.

Unfortunately, no heavy BSM particles have been found in the LHC up to the approximate level of about 1 TeV, with a specific upper limit for each studied process. This seriously changes the situation. Although we can still expect a discovery of a SUSY particle with a mass at the TeV level, the masses of SUSY particles should not be too large. With mass increasing cancellation in the Higgs mass calculations will become less natural and will require a fine-tuning for the parameters of SUSY models. In this situation, the hopes for the SUSY confirmation fade with each new upper limit on the mass of the SUSY particles coming from the LHC.

There are many alternative BSM models, however, a number of renormalizable theories are limited. One such alternative theoretical approach supposes that the Higgs boson is a composite particle with a finite size. Because of that, divergences in the Higgs mass calculations will not appear. The composite Higgs models assume that the Higgs boson has a coupled bound state due to a new type of interaction, which becomes strong around the weak scale. Such models lead to specific predictions of Higgs boson properties. Within this approach, new Higgs-like states will appear at energies of several TeV, which are still not reachable in the LHC. The composite Higgs approach is employed in several specific models, in particular in a rather popular Little Higgs model. There are many other BSM models, in particular the Grand Unification Theory, theories with space extra-dimensions, and so on. The list of different types of BSM models is quite long, however, nobody knows which model is correct.

In the situation of significant uncertainty in the choice of appropriate BSM theory, the experimental search for NP effects becomes the main engine to further advancement. Experimental information on NP effects is required for a critical analysis of existing BSM theories. Therefore, the most important task in particle physics is now to find any evidence of NP effects in collider experiments. There are general theoretical arguments that the lightest NP particle should have a mass less than 1 TeV, or slightly higher, but these arguments are not strictly motivated. Therefore the situation is exciting: the experimentalists expect that NP is just around the corner and they are eager to find it. When NP will be eventually observed it can open up a new horizon for particle physics. Today, there is no known experimental evidence of NP that one can be confident in. When such evidence is found, this will mark the genesis of a new epoch in particle physics. Maybe it will also provide new opportunities for unforeseen technical progress, but we cannot predict it. When the theory of electromagnetism was built, it changed our world completely; however, not all fundamental physics results have practical applications.

1.3 Three frontiers in experimental particle physics

The experimental search for NP phenomena can be conducted along several different paths. In particle physics there are three general directions used to search for NP phenomena: the energy frontier, intensity frontier, and astrophysical frontier. The astrophysical frontier implies the construction of more and more sophisticated apparatuses for astrophysical observations. Recently, this approach was rewarded by an amazing discovery of gravitational waves, which resulted from the fusion of two black holes very far from our Galaxy. Despite a very intensive development of the astrophysics research program in recent decades, this frontier is less appropriate for searches of NP effects because interpretations of astrophysical observations related to particle physics are often ambiguous.

The energy frontier is realized in the construction of particle colliders with a record energy of colliding particles in order to search for the production of new heavy NP particles. In a sense, this is the most direct way to search for NP. The Large Hadron Collider was built with this approach in mind. However, the direct method is not always the most effective. If the masses of new heavy particles are close to several TeV or greater, or their coupling constants are very small, indirect methods of searching for NP should work better.

Indirect BSM searches can be performed in the framework of the intensity frontier. In this approach, it is necessary to build colliders with a record luminosity to study rare processes and look for deviations in their parameters from the SM predictions. This frontier also includes the construction of dedicated complex detectors used to search for rare processes, like neutrino interactions or proton decays. In this frontier, colliders with a very high intensity of beams have to be built to allow for the collection of huge statistics data sets and, thus, to study in detail tiny effects and rare processes. This approach was implemented in the construction of the B factory Belle II, as well as in many other projects. However, in this frontier there are several obstacles; in particular, it requires one to measure tiny effects, which are comparable in size with systematic uncertainties, and very accurate theoretical predictions should be performed.

A certain combination of energy and luminosity frontiers appears in the case of constructing a linear e^+e^- collider, which comprises a high beam energy and the ability to collect large data samples sufficient for a detailed study of heavy particles; in particular, the Higgs boson. It should be noted that the construction of the ILC will allow the investigation of the properties of the Higgs boson with statistics, that is at least an order of magnitude greater than the statistics expected at the LHC experiments. Therefore, the ILC can search for NP effects, which cannot be explored in LHC experiments. Taking into account the fact that the LHC has not yet observed any NP effect, the ILC could become the second most powerful tool to find NP.

1.4 Advantages of e^+e^- colliders

Over the past few years the research program for the ILC has somewhat changed. Before the start of the LHC operation, the ILC was considered as a perfect tool for precise studies of NP particles, which would be discovered at the LHC. It was declared that the LHC is a 'discovery' machine, which should be supplemented by a linear collider called the 'precise measurements' machine.

So far, the LHC has not found any NP particles. The LHC continues to search for new heavy particles, now in the mass range of up to a few TeV. In this situation, the ILC concept has changed. It is still a 'precise measurements' machine, but not for new NP particles. At present, the main goal is precise measurements of the SM heavy particles, namely the Higgs boson, top quark, Z and W bosons. The LHC is a strong competitor to the ILC in this field and we would like to discuss where and why the ILC can make better measurements.

There are several advantages of the ILC in comparison with LHC, mostly due to a difference between the proton–proton and e^+e^- collisions. First of all, recorded events at the ILC are much cleaner than at the LHC. The proton–proton collision cross section is much larger than that for the e^+e^- collisions. In the LHC a serious problem comes from pile-up, when several interactions occur in one beam bunch crossing. The ATLAS and CMS experiments have to operate with approximately 20 collisions on average in each bunch crossing. Because of the many collisions, it is difficult to separate the detector responses corresponding to the studied event. In the ILC, the pile-up is negligible because of a much smaller e^+e^- cross section. In addition, the low collision rate allows for the collection of data without a trigger system, which significantly simplifies data acquisition systems and physics analyses. In the case of a proton–proton collider, additional to a hard interaction of quarks and gluons, the soft interactions of proton remnants will result in an additional underlying-event activity registered in detectors. At the ILC there is no problem with the underlying-event activity because electrons and positrons are not composite.

The second positive feature of the e^+e^- collisions with respect to the pp collisions is the high rate of heavy particles production. In the ILC, the heavy particles will be produced with rates comparable with that for light particles. In particular, at an energy of 250 GeV, the cross section for the Higgs boson production will be ∼300 fb, that is equal approximately to a ∼1% production rate over all e^+e^- collisions. For comparison, at the LHC, the production rate of the Higgs boson is one to one billion pp collisions. At an energy of 380 GeV, the top quark pair production cross section will be ∼650 fb, which is also equivalent to the percentage level among all events. Overall, the high rate of events with heavy particles will simplify physics analyses due to the low backgrounds and high efficiencies of the event selections.

The next advantage of the e^+e^- colliders is the possibility to perform precise theoretical calculations for most of the interesting processes. The colliding electrons and positrons are elementary particles, and their electroweak interaction can be described theoretically with high precision. Radiative corrections are at the level of a few percent and can also be accurately calculated, which allows the reduction of theoretical uncertainties to a one percent level or less. On the contrary, theoretical calculations at the LHC are based on QCD and have large uncertainties up to a few tens of percent for some processes.

The general advantage of the ILC in comparison with other experiments at the intensity frontier is the ability to produce precise measurements of heavy particles, such as the Higgs boson and the top quark. From a theoretical point of view, heavy SM particles should have large couplings with NP particles. Consequently, the most sizable effects of NP can emerge in processes with these heavy particles. There are many promising experiments at the luminosity frontier exploring the intermediate energies, where searches for NP effects are planned. However, the NP effects at the intermediate energies are expected to be very small and it is very difficult to observe them. With some luck, such experiments can win this race for the first observation of an NP manifestation, but the favorites today are the LHC and potentially the ILC.

There is one more point in favor of the ILC: the level of technical preparation of the ILC project is unprecedentedly high at this stage. The R&D for the ILC project has been going on for almost 15 years. To date, all components of the project have been developed in detail, including the technical design of the ILC collider and the ILD and SiD detectors, which are planned to be installed at the ILC. The optimized design and modern technologies are chosen for the collider and detectors, which will allow the reaching of a record accuracy of particle parameters measurements.

Chapter 2

International Linear Collider project technical aspects

Tremendous progress has been achieved in particle physics during the last decades. To a large extent, this was caused by the fast development of particle acceleration technologies. The beam energy of accelerators and, later, colliders grew almost exponentially with a year of construction. The fast development of acceleration technologies, along with the hard and very productive work of many experimentalists and theorists, allowed us to penetrate incredibly deep into the secrets of the microworld. It should be noted that the progress in particle physics was achieved in particular due to the very attractive feature of collider experiments: measurements can be repeated and tested with larger statistics. A new collider can be built and obtained earlier experimental results can be critically tested. It provides the possibility to rule out wrong results quickly and very effectively.

Now the exponential tendency of the collider energies rise is exhausted. This happened because, similar to the collider energy, project costs and the number of scientists involved also increased almost exponentially. These resources have certain limitations and we are now close to these limits. There are proposals to construct a hadron machine with an energy of 50–100 TeV in CERN or in China, near Beijing, in 20–30 years. However, such projects are in a very early stage of discussion and will require tens of billions of US dollars. In this situation, it is reasonable to now start the construction of a linear collider, which can be built and start operations in about 10 years. Possible alternatives to the linear collider project could include the construction of a machine based on new principles; like a $\mu^+\mu^-$ collider, or maybe a collider using new methods of the particle acceleration, such as the plasma acceleration technique. Unfortunately, both these directions seem to require a long R&D period of about 20–30 years to come to the possibility of a practical realization.

doi:10.1088/2053-2571/aae221ch2

2.1 International Linear Collider design

The concept of the e^+e^- linear collider was proposed for the first time in 1965. Since the 1990s, the practical realization of the project began to be discussed. Several e^+e^- linear collider projects were proposed, in particular SLC and NLC in USA, JLC in Japan, and the TESLA project in Germany. Unfortunately, none of these projects were approved. If approved, such a project could become the first to discover the Higgs boson.

The International Linear Collider (ILC) project in Japan was starting to be considered almost 15 years ago. The technical parameters of the collider were first outlined in 2003 and later updated in 2006. The ILC project was further developed and all relevant information was collected and published in 2013 in the Technical Design Report (TDR). The report included a detailed description of the R&D results obtained in three main areas: the ILC design, the design of two detectors that are planned to be installed at ILC, the International Large Detector (ILD) and the Silicon Detector (SiD), and a physics research program.

The high-luminosity e^+e^- linear collider with the center-of-mass energies in the range (200–500) GeV was presented in the TDR with a possible extension up to 1 TeV. Initially the collider length was about 31 km and the maximum energy of 500 GeV was assumed. The possibility of increasing the length of the collider up to 50 km was planned in order to achieve an energy of 1 TeV. The final electron and positron acceleration was based on 1.3 GHz superconducting radio frequency (RF) accelerating technology. The collider luminosity was planned to achieve 2×10^{34} cm^{-2} s^{-1}. The stability of the beam energy was expected to be within 0.1%. According to the design, both beams will be polarized, P(e^-, e^+) = (−0.8, +0.3). In the TDR, the approximate cost of the project was estimated to be ~8 billion US dollars.

Recently, the Japan Association of High Energy Physicists (JAHEP) proposed the building of the ILC collider in stages, with the center-of-mass energy in the first stage limited to 250 GeV. The proposal was motivated by two reasons: the main goal of the ILC is now the precise Higgs studies, which can be performed at 250 GeV, and the 250 GeV machine will be much cheaper. The length of the 250 GeV collider can be reduced to ~20 km and the cost will be less by about 30%. Despite the length decrease, the overall design of the 250 GeV collider complex remains almost the same as was proposed in the TDR for the 500 GeV machine. The schematic layout of the ILC, taken from the TDR, is shown in figure 2.1. In the TDR, the average final acceleration gradient was 31.5 MV/m and the pulse length was 1.6 ms. In the latest version of the technical design, it is assumed that the gradient can be increased to 35 MV/m.

As shown in figure 2.1, the electron beam is produced in the central region of the accelerator tunnel by a laser illuminating the GaAs photocathode. The electrons are emitted with 90% polarization. Their beams are formed and injected into the electron damping ring. The positron beam is obtained in a different method using the photoproduction process induced by the electron beam. Photons with energies in the (10–30) MeV range are produced when the electron beam is transported through a 147 m specially built-in superconducting helical undulator. The photons are directed at a Ti-alloy target located ~500 m downstream, where they produce

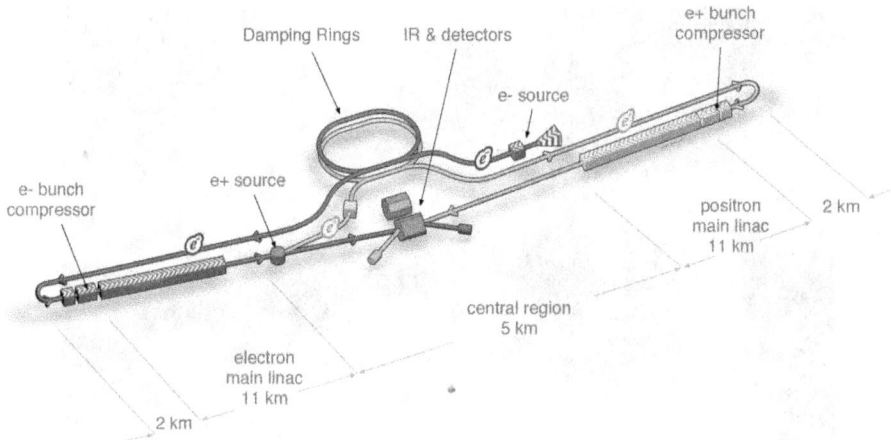

Figure 2.1. Schematic view of the ILC accelerator complex is shown with all major subsystems included. Source: ILC TDR Volume 1, arXiv:1306.6327, figure 3.1.

electron–positron pairs. Then, the positrons are separated from the electrons and photons, polarized, 'cooled', and injected into the positron damping ring. In the baseline design, a positron polarization of 30% is planned, however, a larger polarization is technically possible with a longer undulator.

The electrons and positrons will circulate in the storage rings with the energy of ~5 GeV. Then, the formed electron and positron beams enter the transportation system to the linac accelerator. This system consists of a 15 km part that focuses the particles with 5 GeV of energy and delivers them to the region of a turn for 180°. After the turn, the electrons and positrons are accelerated from 5 GeV to 15 GeV, and then directed to the main linac system, where their final acceleration is performed. A key component that accelerates particles to the final energies is the 11 km main linac system, which is based on superconducting RF acceleration technology and consists of a number of niobium cavities (figure 2.2). Two similar linac systems are planned to accelerate electrons and positrons from 15 GeV to the final collision energy between 100 GeV and 250 GeV. Since the beam energies will be limited to 125 GeV at the first stage, the length of the main linac system will be reduced from 11 km to ~6 km for each beam. After acceleration, the electrons and positrons enter the final beam delivery system, where they will be additionally focused and will collide in the center of an installed detector.

In the collision region, the beams are very small in size, which makes it possible to achieve a high luminosity. The vertical size of the beams in the collision region is (5–8) nm, the horizontal size is about 0.7 μm. The beams will collide with the crossing angle of 14 mrad. The planned beam currents should be ~(5–10) mA. The expected frequency of the e^+e^- collisions is of the order of (5–10) Hz, which allows one to record data without a trigger system. The beams coming out of the collision area will be scattered and directed into a dumping system.

Polarization of the electron and positron beams provides additional opportunities for experimental measurements, since the specific Standard Model and New Physics

Figure 2.2. The niobium cavities are the key element of the linac system. The required production accuracy of the cavity geometry is ∼1 μm. © Fermilab.

processes can be enhanced by choosing an appropriate beam polarization. At the moment it is planned to start data taking with the beam polarization of P $(e^-, e^+) =$ $(-0.8, +0.3)$, with a possibility for switching to the opposite polarization P $(e^-, e^+) =$ $(+0.8, -0.3)$. It should be noted that electron–positron annihilation is possible only when the colliding electrons and positrons have different helicities, $e_L^- e_R^+$ or $e_R^- e_L^+$. Due to this, the oppositely polarized beams allow one to increase the luminosity.

Various options of data taking were discussed in the TDR, where a few hundred fb^{-1} of integrated luminosity were planned to be collected at each center-of-mass energy of 250 GeV, 350 GeV, and 500 GeV. When the staged approach was proposed, it was decided to stay at 250 GeV for a long period of time in order to collect the total luminosity of about 2 ab^{-1} (=2000 fb^{-1}). This amount of data will be assumed below in estimates of the significance of signals for studied physical processes.

2.2 ILD detector design

The ILD and SiD detectors are planned to be installed in the ILC. These detectors were designed by two independent collaborations, and corresponding R&D studies have been going on for many years. Unfortunately, there is only one interaction point on the ILC, where the detectors can be installed. Therefore, these two detectors are planned to share the operation time in the beam position. In this book, only the ILD detector will be discussed in detail. The SiD detector has a slightly different concept, this is a compact detector with a 5 Tesla magnetic field and silicon vertex and tracking subsystems. It is possible that, should the ILC project be approved, it will be decided to build only one detector. Then, the subdetectors technologies developed by both collaborations will be compared and the best options will be chosen or maybe somehow combined in one detector.

The detectors, proposed for the ILC, differ significantly from the detectors ATLAS and CMS, installed at the Large Hadron Collider (LHC). It is because the backgrounds and collision rates are significantly lower in the e^+e^- collisions than those in the pp collisions. The ILD detector was developed taking into account the energies of the final state particles and the low e^+e^- collision rate. The energy range of particles recorded in the ILD up to a few hundred GeV allows one to build a relatively compact detector, which is much smaller than the ATLAS and CMS detectors, however, larger than the SiD detector. The low collision rate provides the opportunity to use specific detection technologies, which are difficult to use for the ATLAS and CMS detectors because of a much larger collision rate. Over the years, after the LHC construction, a number of new modern technologies appeared, which are applied in the ILD detector design. Due to these reasons, the ILD detector was designed with the goal of achieving a record accuracy of measuring the coordinates of secondary vertices, charged particle momenta and the energies of jets.

The general view of the detector ILD on the platform and the subdetectors positioning in the quadrant are shown in figure 2.3. The ILD detector is relatively small in size, hermetic, and covers a large angular interval. The main emphasis in the development of the detector components was placed on high accuracy of particle parameter measurements. During the last years, particle flow algorithm (PFA) was developed, which allows for significant improvement in the accuracy of the jet energy measurements. The PFA method works well in the case of calorimeters with small cells. The ILD detector was designed to use this method in full scale.

As you can see in figure 2.3, the vertex detector is located the closest to the interaction point. Multi-layer pixel semiconductor technology is used for this subdetector. The main purpose of the vertex detector is a precise measurement of secondary decay vertices, corresponding to weak decays of long-lived particles, like B and D mesons, or τ leptons. The distance between the primary decay vertex and the secondary decay vertices can vary from zero to a few cm and it is important to

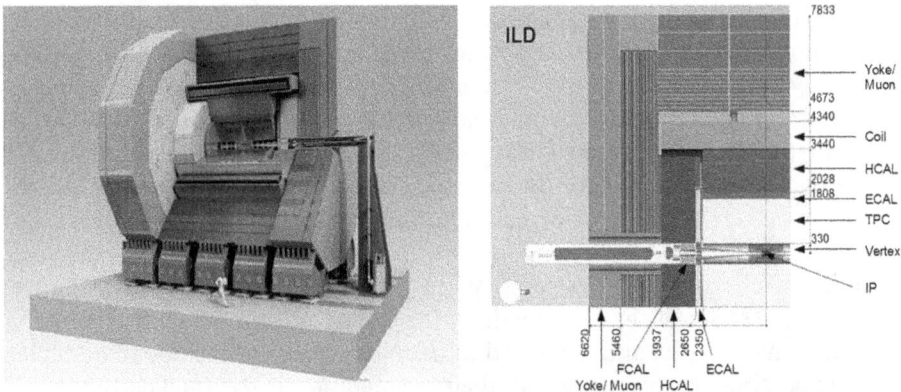

Figure 2.3. The isometric view of the ILD detector (left) and the subdetectors positioning in the quadrant (right). Source: ILC TDR Volume 4, arXiv:1306.6329, figures III.1.1 and III.1.2.

have high accuracy in the vertex coordinates measurement. The spatial resolution of the secondary vertex with the proposed technology will be as good as (3–5) μm. Moreover, using information from the vertex detector, the accuracy of charged track momenta measurements obtained in the main track system can be slightly improved. The vertex detector must tolerate the collider radiation and the proposed technology must have a high radiation resistance.

Measurement of the momenta of charged tracks will be performed by a hybrid tracking system, including a few layers of silicon strips and the time projection chamber (TPC). The system is located just after the vertex detector and occupies a relatively large volume in the detector. To obtain a good momentum resolution, the subdetector will be situated in a strong magnetic field of 3.5 T. The used TPC technology is based on a drift of ionization clusters, produced by charged particles in a homogeneous electric field to the chamber wall. The track coordinates x and y are measured from the cluster projection on the chamber wall equipped by end-plate detectors, and the coordinate z is measured from the drift time. A large number (up to 224) of the cluster coordinates will be measured with the accuracy of the order of 100 μm for all three coordinates. The whole tracking system allows reaching a very high momentum resolution of $\sigma_{1/p_T} = 2 \times 10^{-5}$ GeV^{-1}. The main TPC volume is filled with gas. This means a very small amount of material, which results in a small impact on particles coming through the TPC and registered in calorimeters. This is important for PFA method application, and also allows one to minimize the effects due to the high collider radiation penetrating the TPC. Another advantage of the TPC is a high reconstruction efficiency, which will be close to 100% for tracks with a momentum larger than 1 GeV. The disadvantage of the TPC is its large dead time, however, this is not critical for the e^+e^- collider experiment with the low collision rates. This characteristic makes it difficult to use TPC technology in high collision rate ATLAS and CMS experiments.

The tracker is surrounded by an electromagnetic calorimeter (ECAL) and, further, by a hadronic calorimeter (HCAL). The ILD calorimeters have very high granularity, which provides a record jet energy resolution. However, it requires having many millions of scintillator cells. The high ECAL granularity allows one to effectively resolve two photons emitted with close angles. The small size of the HCAL calorimeter cells is very important when applying the PFA method, which significantly improves the energy resolution of hadronic jets. With the planned granularity, the jet energy resolution is expected to achieve $\sigma_E / E \sim (3–4)\%$, which is equivalent to $30\% / \sqrt{E}$ at 100 GeV. Such excellent jet energy resolution makes possible the separation of Z and W bosons in hadronic decay modes.

In the basic version of the hadronic calorimeter, the cell size is 3×3 cm^2, which allows in many cases the tracing of the individual trajectory and the energy deposition of charged particles, even when they are produced inside a hadronic jet. In the framework of the PFA method, the energy of the charged particles flying from the primary vertex and measured in the hadronic calorimeter can be replaced by the value obtained by measuring their momentum in the tracker system. This replacement allows one to significantly improve the jet energy resolution.

Immediately after the calorimeter, a coil of superconducting magnets will be positioned, providing the axial magnetic field of 3.5 T in the inner part of the detector. Next after the coil the iron yoke of the magnet is located, which is equipped with scintillation strips or planar resistive detectors, which will provide a muon identification. Currently, the final choice between these two technologies for the muon subdetector is not made. It has to be noted, that a potential set of options also still exists for other subsystems.

One of the important issues of event reconstruction at linear collider detectors is an accurate account of the radiation effects associated with electron and positron beams. It includes photon radiation in the initial state, photon radiation in the final state, and the emission of photons by the beam particles of one of the colliding bunches in the electromagnetic field of another bunch (beamstrahlung). These effects can significantly distort studied signals and lead to additional background activity in the detector, especially in the areas close to the beam pipe. The direction of radiative photons is usually close to the direction of the emitting electron or positron. Special attention will be paid to the shielding of the beamstrahlung photons, and also to accurate simulation of these processes in the detector. This will require a detailed knowledge of the beam configurations in the collision area. Computer codes were developed, which describe these three photon radiation processes, and were used for detector optimization, based on simulation of benchmark processes.

2.3 Alternative e^+e^- collider projects

A long time ago, another linear e^+e^- collider project, the Compact Linear Collider (CLIC), with energy in the range (380–3000) GeV was proposed at CERN. The collider is schematically shown in figure 2.4. The CLIC conceptual design report (CDR) was published in 2012. Although the designs of the CLIC and ILC detectors are very similar, their accceleration systems are based on different concepts. The final acceleration gradient in the CLIC is planned to be 100 MV/m, which is approximately three times larger than in the ILC. The normal-conducting two-beam acceleration scheme is used to fulfill this condition. Details of the method can be found in the CLIC CDR published in 2012. The high acceleration gradient is critical to reach multi-TeV energy regions in a reasonable collider length. Recently, CLIC collaboration has also decided to apply a staging approach, with the 380 GeV center-of-mass energy at the first stage. The energy was chosen with a goal to collect a large data sample of top quarks because the top–antitop pairs will be numerously produced at this energy. The Higgs boson production cross section at 380 GeV is about the same as at 250 GeV, resulting in a similar number of reconstructed Higgs bosons. At the first stage, the length of the collider is planned to be about 10 km, and the cost of the project was estimated at about 7 billion Swiss francs.

In 2012, the circular electron positron collider (CepC) project was proposed in China. In the CepC, the electron and positron beams will circulate in rings with an energy of 120 GeV. The collider energy is limited since the beam radiation rapidly increases with energy because of the beam bending. Now the CepC CDR is under preparation. In a circular collider, two or more beam interaction points are possible

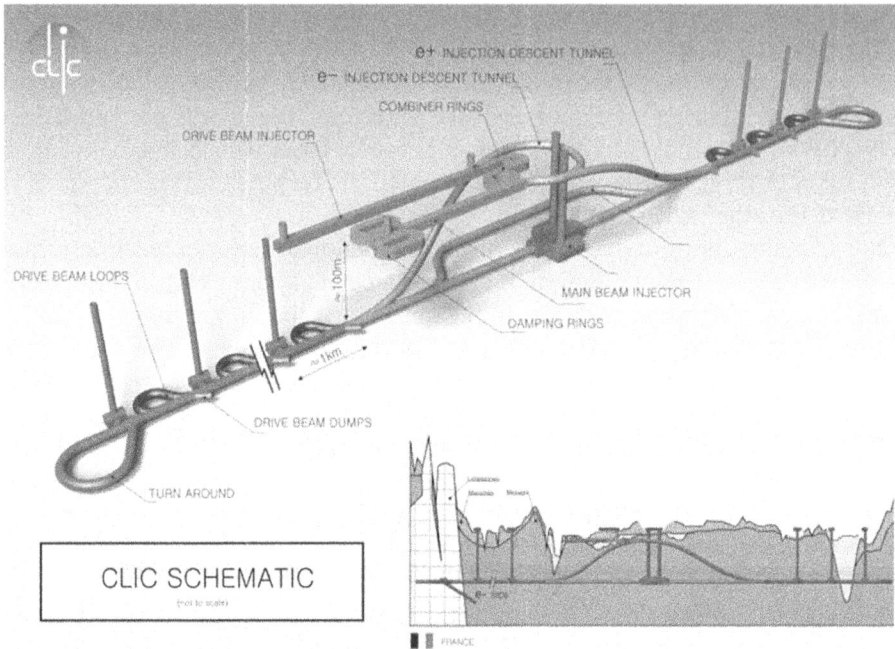

Figure 2.4. Schematic view of the CLIC accelerator complex. The layout below shows how the collider is located underground relative to the Earth surface.

and, accordingly, two detectors are planned. It was recently proposed to increase the length of the circle from 54 to 100 km, but the final length has not yet been fixed. It is expected that the construction of the CepC will begin in 2022 and be completed in 2030. It is interesting to note that the planned luminosities of all three colliders, the ILC, CLIC and CepC, are rather similar and have to reach about 2×10^{34} cm^{-2} s^{-1} at their full operation regime. The designs of the detectors at these three projections are also quite similar. Recently a circular e^+e^- collider FCC-ee has begun to be discussed at CERN. The centre-of-mass energy of the FCC-ee collider is planned to vary from 90 to 400 GeV. This FCC-ee project is part and parcel of the Future Circular Collider (FCC) design study at CERN, and would be the first step towards the long-term goal of a 100 TeV proton–proton collider. The collider will be located in a new 80–100 km tunnel in the Geneva area.

In addition to the basic regime of the linear e^+e^- colliders, there are a number of interesting ideas to using other operation modes. In particular, it is possible to take data at the Z boson mass peak of about 90 GeV to collect a Z boson dataset, which will be a thousand times larger than was obtained in the LEP experiments. This opportunity was discussed at the beginning of the ILC project preparation, but was later removed from the agenda due to a lack of resources. There is also an idea to upgrade the e^+e^- collider to the $\gamma\gamma$ or $e^-\gamma$ colliders. Technically, it can be realized with the help of powerful short-pulsed lasers producing photons, which will be scattered by electrons with a large momentum transfer. The $\gamma\gamma$ collider has a great physics potential for performing a number of important and actual measurements. This option has also not been developed due to limited resources. The potential for possible modifications of the e^+e^- colliders is obvious and will undoubtedly be realized in the future.

Chapter 3

Physical research program at the International Linear Collider

The main topic of research for the International Linear Collider (ILC), operating at 250 GeV, will be the study of the Higgs boson. Of course, measurements of the Higgs boson parameters can also be performed at a higher energy. In contrast, the top quark can be studied only at energies larger than 350 GeV, where the top-pair production channel will open. At a higher energy, exceeding the top-pair mass threshold, the studies of the top quark will become the second main research topic. In this book we will concentrate on the physics case at 250 GeV, however, physics at higher energies will also be discussed.

In this book, the Higgs boson predicted in the Standard Model (SM) and observed at CERN with a mass of $(125.09 \pm 0.21 \pm 0.11)$ GeV will be denoted as H. It should be noted that in some theoretical papers this 'light' Higgs boson is denoted as h, whereas a heavy Higgs boson predicted in some BSM models is designated as H. These light and heavy BSM Higgs bosons will be discussed below, they will be denoted as H_1 and H_2, respectively.

At 250 GeV, the Higgs boson will be produced mainly in the channel

$$e^+e^- \rightarrow Z\,H \qquad (3.1)$$

with a cross section of $\sigma \approx 300$ fb. According to the plans, a data sample with the integrated luminosity $\mathcal{L} \approx 2000$ fb^{-1} will be collected at the ILC during the time of operation, and this sample will comprise about six hundred thousand Higgs bosons:

$$N(H) = \sigma(e^+e^- \rightarrow Z\,H) \times \mathcal{L} = 6 \cdot 10^5. \qquad (3.2)$$

Because of such numerous productions of the Higgs bosons, the ILC is often called a 'Higgs factory'. Taking into account the high reconstruction efficiency of the Higgs boson at ILC, the number of reconstructed Higgs bosons is expected to be at least an order of magnitude larger than is expected in the LHC experiments even after the luminosity upgrade.

In a sense, when a large statistics is collected, the 125 GeV Higgs boson can serve as a 'window' in New Physics (NP). We can see NP effects indirectly in this 'window', using measured parameters of the Higgs boson. If these parameters are affected by NP contributions, they will differ from those predicted in the framework of the SM. Therefore, it is very important to perform precise measurements of Higgs boson parameters. One of the main topics in experimental studies of the Higgs boson at 250 GeV is the measurements of the production cross sections, and the decay branching fractions of the Higgs boson. The experimental technique used in the ILC allows one to measure the parameters in a model-independent way. In addition, it is possible to measure the decay branching fraction in the invisible final state. These measurements will allow one to obtain the values of the Higgs boson couplings with leptons, quarks, and vector bosons with high accuracy. The measurements are very important because the values of the couplings are sensitive to NP effects.

The mass and width of the Higgs boson can be measured with an accuracy significantly better than the current world records. The CP content of the 125 GeV Higgs boson can be tested. It is important to set a restrictive upper limit on the CP-odd admixture or maybe to observe this component in the 125 GeV Higgs boson, since the SM Higgs boson is predicted to be 100% CP-even. Taking into account the expected large number of Higgs bosons, the high reconstruction efficiency, and low backgrounds, rare decays of the Higgs boson can be searched for.

Unfortunately, the pair production of top quarks is possible only at energies above 350 GeV. If these energies are reached at the next ILC stage, it will be possible to measure the mass and width of the top quark with uncertainties that are an order of magnitude better than those obtained now. These top quark parameters are sensitive to BSM effects. The current world best accuracy for these parameters is not sufficient for critical tests of the SM.

An important task is to measure the $t\bar{t}Z$ and $t\bar{t}\gamma$ couplings, which can deviate from SM predictions in different BSM models. A large number of top quarks expected in a collected data sample will allow the search for rare top quark decays. Angular distributions of top quark decay products can contain important information, since in previous experiments there were small discrepancies with the SM in such measurements. An additional opportunity occurs at the ILC at 250 GeV, where there is the possibility of a precise measuring of the parameters of the gauge W and Z bosons. It is also possible at the ILC to search for direct production of the NP particles with anomalously small cross sections, which are difficult to detect in the LHC.

3.1 Higgs boson studies

3.1.1 Higgs boson theoretical description in the SM

In the SM the most general SU(2) × U(1)-invariant renormalizable Higgs potential is given by:

$$V(\Phi) = m^2|\Phi|^2 + \lambda|\Phi|^4 \tag{3.3}$$

where the constant λ is positive and $m^2 < 0$. The electroweak symmetry is spontaneously broken in this potential, generating masses of W and Z bosons. The Higgs field Φ is a complex isospin doublet with a hypercharge $Y = 1$, which is parameterized as:

$$\Phi = \frac{1}{\sqrt{2}}\begin{pmatrix} \sqrt{2}\,\phi^+ \\ \phi^0 + ia^0 \end{pmatrix}, \tag{3.4}$$

where ϕ^+ is the complex charged component of the Higgs doublet, and ϕ^0 and a^0 are the CP-even and CP-odd neutral components, respectively. The component ϕ^0 is the physical Higgs boson state. The neutral a^0 and charged ϕ^+ Goldstone boson components are absorbed by the longitudinal components of the Z, W^+ and W^- physical gauge bosons. As a result, only one neutral Higgs boson with the quantum numbers $J^{CP} = 0^{++}$ is predicted in the SM.

The Lagrangian describing the interaction of the Higgs boson with fermions, gauge bosons, and the Higgs boson itself can be represented in the form

$$\mathcal{L} = -g_{Hf\bar{f}}\, H\bar{f}f + \delta_V g_{HVV}\, H\, V_\mu V^\mu + 1/6 \times g_{HHH}\, H^3 \tag{3.5}$$

where $V = W^\pm(Z^0)$, $\delta_V = 1$ (for $V = W^\pm$) or $1/2$ (for $V = Z^0$). The values of the couplings are proportional to the corresponding fermion mass or to the square of the boson mass:

$$g_{Hf\bar{f}} = m_f/v, \qquad g_{HVV} = 2m_V^2/v, \qquad g_{HHH} = 3m_H^2/v. \tag{3.6}$$

The vacuum expectation value of the Higgs doublet is $v = (\sqrt{2}\,G_F)^{-1/2} \approx 246$ GeV, which can be determined from the Fermi constant with high accuracy. Proportionality of the couplings to the fermion masses and the square of the masses of the Z, W and Higgs bosons is the basic prediction of the SM. Any deviation from proportionality will be a direct manifestation of NP.

It is difficult to include neutrinos in the Higgs boson theory within the framework of the SM. As it is treated today, the right-handed neutrino fields are absent, therefore, the Higgs boson cannot couple with neutrinos. Consequently, the Higgs boson mechanism cannot generate the neutrino masses, which are observed experimentally. However, the very small neutrino masses have almost no impact on Higgs physics. In a reasonable approximation, the neutrinos can be simply treated as massless.

If the mass of the Higgs boson is precisely known, the couplings and, respectively, the decay branching fractions of the Higgs boson can be calculated in the framework

Table 3.1. The theoretical calculations of the decay branching fractions (in %) and their theoretical uncertainties (in relative %) for the main Higgs boson decay channels for the mass $m(H) = 125.09$ GeV.

$b\bar{b}$	$c\bar{c}$	$\tau^+\tau^-$	$\mu^+\mu^-$	WW^*	ZZ^*	gg	$\gamma\gamma$	$Z\gamma$
58.1	2.88	6.26	0.0217	21.5	2.64	8.18	0.227	0.154
$\pm 1.3\%$	$^{+5.5}_{-2.0}\%$	$\pm 1.6\%$	$\pm 1.7\%$	$\pm 1.5\%$	$\pm 1.5\%$	$\pm 5.2\%$	$\pm 2.1\%$	$\pm 5.8\%$

of the SM (see table 3.1). In the calculations of these branching fractions, the corrections are small and can be accurately estimated.

In the so-called κ-formalism, the couplings g_i are parametrized as its ratios to their SM values:

$$\kappa_i = g_i \,/\, g_i^{\text{SM}}. \tag{3.7}$$

If the experimental results agree with the SM, all measured modifiers κ_i must be equal to the unit. In the BSM models, the modifiers κ_i may deviate from the unit values.

In the leading order, the decays of the Higgs boson into fermions (figure 3.1(a)) and vector bosons (figure 3.1(b)) are described by 'tree' diagrams. It should be noted that in the decays $H \rightarrow WW^*$ and $H \rightarrow ZZ^*$ at least one vector boson has an off-shell mass. The Higgs boson self-coupling is an important parameter of the SM. The strength of the Higgs boson self-couplings is proportional to the square of its mass. The observed Higgs mass of 125 GeV implies that the Higgs dynamics corresponds to the weak scale.

The Higgs boson cannot couple to photons and gluons by the 'tree' diagram. This is because the Higgs boson has no charge or color. Therefore, the decays of the Higgs boson into two gluons, two photons, and $Z\gamma$ can only be described by the 'loop' diagrams. In the SM, the effective $H\gamma\gamma$ and $HZ\gamma$ couplings are induced via the top loop (figure 3.1(c)) and the W loop (figure 3.1(d)) diagrams, where the W loop diagram dominates. The Hgg coupling is induced by the quark loop diagrams with a dominant contribution from the top quark loop. Usually the processes described by the loop diagrams are suppressed with respect to those described by the tree diagrams.

From the measurements of the Higgs boson decays in the gg, $\gamma\gamma$ and $Z\gamma$ channels, the effective modifiers κ_g, κ_γ, and $\kappa_{Z\gamma}$, respectively, can be obtained. Then the modifiers κ_W and κ_t of the g_{HWW} and $g_{Ht\bar{t}}$ couplings can be estimated indirectly using the following formulas:

$$\begin{aligned}
\kappa_g^2 &= 1.06\,\kappa_t^2 - 0.07\kappa_t\kappa_b + 0.01\kappa_b^2 \\
\kappa_\gamma^2 &= 1.59\kappa_W^2 - 0.66\kappa_W\kappa_t + 0.07\kappa_t^2 \\
\kappa_{Z\gamma}^2 &= 71.12\kappa_W^2 - 0.15\kappa_W\kappa_t + 0.03\kappa_t^2
\end{aligned} \tag{3.8}$$

Here the modifier κ_b corresponds to the $g_{Hb\bar{b}}$ coupling.

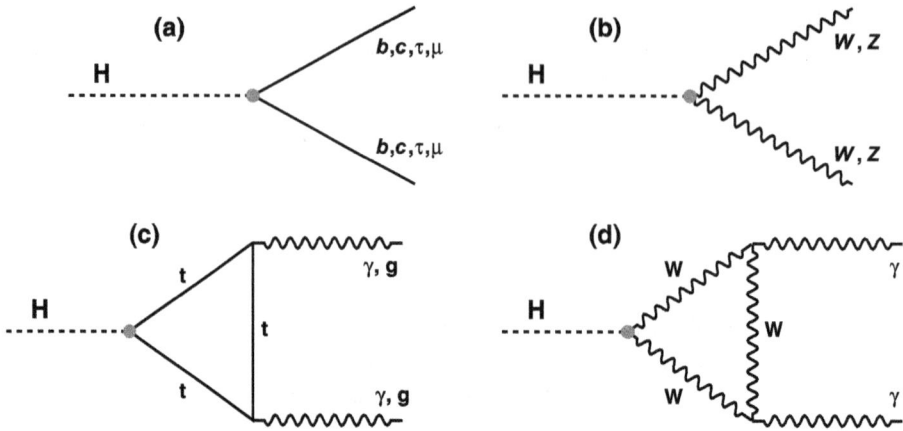

Figure 3.1. The leading order diagrams for the Higgs boson decays are shown: 'tree' diagrams for decays to (a) fermions and (b) gauge bosons and the (c) t 'loop' and (d) W 'loop' diagrams for the final state $\gamma\gamma$. The Higgs boson couplings are marked by red dots.

The mass of the Higgs boson is not predicted in the SM. As mentioned above, it is difficult to explain the low Higgs boson mass in the framework of the SM. The width of the Higgs boson Γ_H can be predicted in the SM with high accuracy, if its mass is fixed. For the mass $m(H) = 125.1$ GeV the width should be (4.1 ± 0.1) MeV.

3.1.2 BSM models with extended Higgs sector

There are many BSM models, which include more than one Higgs boson. In some of these models, additional Higgs doublets are introduced without other modifications to the SM. Initially, this approach was motivated by attempts to find a new source of the *CP*-violation. An example of such a type of model is the Two Higgs Doublet Model (2HDM). The Higgs sector can also be extended in a more complex Minimum Supersymmetric Standard Model (MSSM), where the superpartners to all SM particles are introduced, and additional Higgs doublets appear naturally as a consequence of a more general approach.

The simplest and most popular Higgs sector extension scheme is presented in the 2HDM Type II model, which coincides with a basic version of the MSSM model from the point of view of the Higgs sector. In the framework of this approach a second Higgs doublet is added. The form of the second doublet is similar to the SM doublet given in equation (3.4), however all three components of the second doublet are represented by physical states, resulting in the four new Higgs fields, in addition to one Higgs field from the first doublet. These five physical Higgs bosons include two *CP*-even scalar bosons, a 'light' H_1 and a 'heavy' H_2, one *CP*-odd boson A^0, and two charged bosons $H^{\pm}$. The light boson H_1 should have properties similar to the SM Higgs boson H, which is identical to the Higgs boson with a mass of ~125 GeV discovered at CERN. In some types of these models, the three neutral Higgs bosons, H_1, H_2 and A^0, can mix, which should lead to the appearance of the *CP*-odd component in the Higgs boson with a mass of 125 GeV. Within the general 2HDM approach, there are several types of models, which differ in details.

There are extensions of the Higgs sector with a more complex structure. In particular, the Higgs Triplet Model was proposed, in which many Higgs fields will appear, including doubly charged Higgs bosons $H^{\pm\pm}$. In all the models discussed above there are additional heavy Higgs bosons.

Potentially, if the new Higgs bosons are not too heavy, their couplings are weak, and the energy of a e^+e^- collider is large enough, these bosons could be observed in the processes:

$$
\begin{aligned}
&\text{(a) } e^+e^- \to H_1\, A^0 \\
&\text{(b) } e^+e^- \to H^+H^- \\
&\text{(c) } e^+e^- \to H^\pm W^\mp.
\end{aligned}
\tag{3.9}
$$

It is interesting to note that the process (c) in equation (3.9) is strongly suppressed in 2HDM models, however this process is allowed in extended Higgs sector models with three or more doublets. Because new Higgs bosons have not been observed at LHC experiments, the chances that they will be directly produced and registered in the ILC are very low. However, a reflection of these heavy Higgs bosons can be found in the ILC indirectly, as will be explained below.

As an alternative to point-like Higgs models, the composite Higgs models were proposed. Within the framework of these models, new strong interactions are introduced on a scale of the order of 10 TeV. Typical examples of this approach are the Little Higgs model and the theories formulated in a space of Higher Dimensions. In the basic version of the Little Higgs model, ten Higgs bosons will appear with masses in the region of several TeV. An analog of the SM Higgs boson with a mass of 125 GeV will emerge from a degenerate Higgs doublet. Doubly charged Higgs bosons can also exist in the Little Higgs model. As a consequence of this model, the couplings of the 125 GeV Higgs boson will be affected at a level around several percent.

There are specific predictions in all versions of these BSM models with extended Higgs sector, which can be tested experimentally. First of all, additional Higgs bosons will lead to a small modification in the measured values of the SM Higgs boson couplings as a result of additional contributions via 'tree' and 'loop' diagrams. Within the supersymmetric (SUSY) models, extra contributions to the couplings can also come from 'loop' diagrams, which include in the loop other SUSY particles, like stop quarks.

The main purpose of couplings measuring is to find deviations of the obtained values of the Higgs boson couplings from SM predictions. Each BSM model has a specific pattern of coupling deviations. Such 'fingerprints' in the chart of the coupling deviations for the Higgs boson with a mass of 125 GeV from the values of the SM for different models of the NP are shown in figure 3.2. As one can see, the deviations from the SM values can be distinguished in different BSM models, because they are larger than the experimental uncertainties expected at ILC. As can be seen from these figures, the deviations can reach $\sim(5\text{–}10)\%$ for certain channels. This means that, for the critical testing of the BSM models, the experimental uncertainties must be of about 1% or less. As will be shown below, it is possible in the ILC.

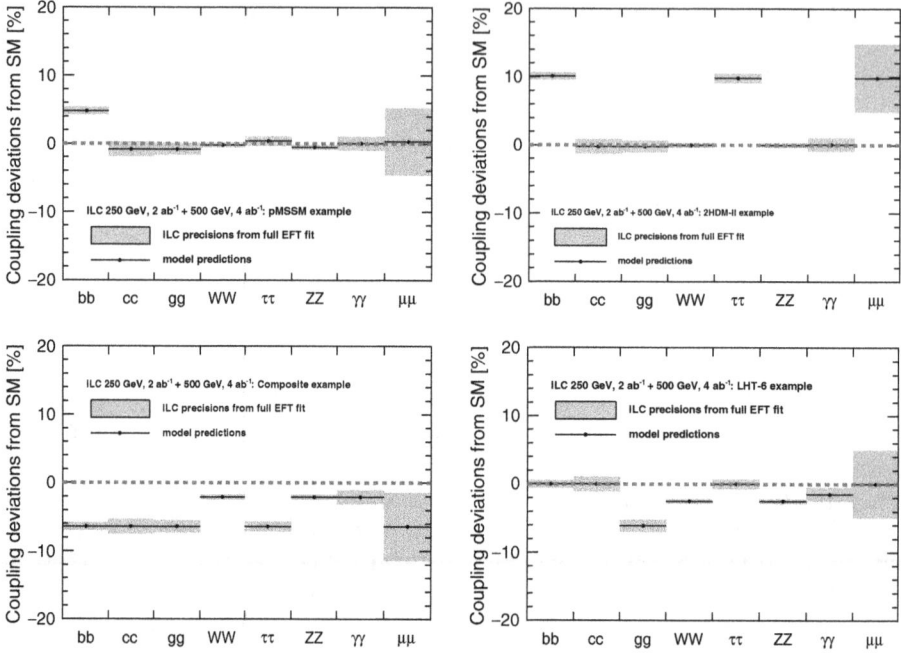

Figure 3.2. The deviations of the Higgs boson couplings from the Standard Model predictions (in percent, black lines) are compared with the uncertainties of the expected experimental measurement (pink belts) in the ILC. The charts are shown for specific BSM models: the MSSM model (top left), the 2HDM Type-II model (top right), the Composite Higgs model (bottom left) and the Little Higgs model (bottom right). Source: Barklow *et al* 2018 Phys. Rev. **D 97** 053003, figure 3.

In addition to the coupling measurements, there are a number of other parameters that can indicate the effects of NP. In particular, the observation of the *CP*-odd component in the Higgs boson will be a direct manifestation of an NP contribution. Therefore, it is necessary to obtain a strict upper limit on this parameter, or maybe to find the *CP*-odd component contribution. Branching fractions for various rare decays of the Higgs boson should be measured and compared with SM predictions. Also, the branching fraction of the Higgs boson decay in the invisible final state should be precisely measured in order to search for a contribution from new unknown particles. Any deviation from the SM predictions will be direct evidence of an NP contribution.

3.1.3 Higgs boson production processes

For the ILC operating at 250 GeV, the Higgs boson will be produced in three processes:

$$
\begin{aligned}
&\text{(a)} \quad \text{Higgsstrahlung:} \; e^+e^- \to H\,Z \\
&\text{(b)} \quad WW \text{ fusion:} \quad e^+e^- \to H\,\nu_e\,\bar{\nu}_e \\
&\text{(c)} \quad ZZ \text{ fusion:} \quad e^+e^- \to He^+e^-.
\end{aligned}
\tag{3.10}
$$

The 'tree' diagrams of these processes are shown in figure 3.3. The energy dependence of the cross sections for these channels is shown in figure 3.4. As we can see, at 250 GeV the Higgsstrahlung process (a) will be dominant with a cross section of about 300 fb. The WW fusion process (b) will have a cross section of about 10–15 times less, and the ZZ fusion process (c) is strongly suppressed. With increasing energy, the cross section (a) decreases, whereas the cross sections (b) and (c) rise. At ~450 GeV, the cross sections (a) and (b) will be approximately the same.

It has to be noted that the beams polarization $P(e^-, e^+) = (-0.8, +0.3)$ was assumed in the calculations of the cross sections shown in figure 3.4. The oppositely polarised beams lead to an increase in the luminosity. It is important that the ILC will have the technical ability to change the beam polarizations, in particular, to the

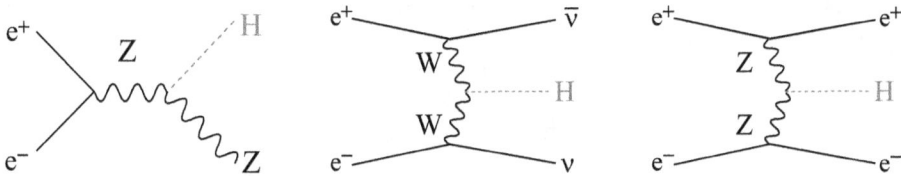

Figure 3.3. Diagrams describing processes (from left to right): (a) $e^+e^- \rightarrow H Z$, (b) $e^+e^- \rightarrow H \nu_e \bar{\nu}_e$ and (c) $e^+e^- \rightarrow He^+e^-$. The Higgs boson is shown in red.

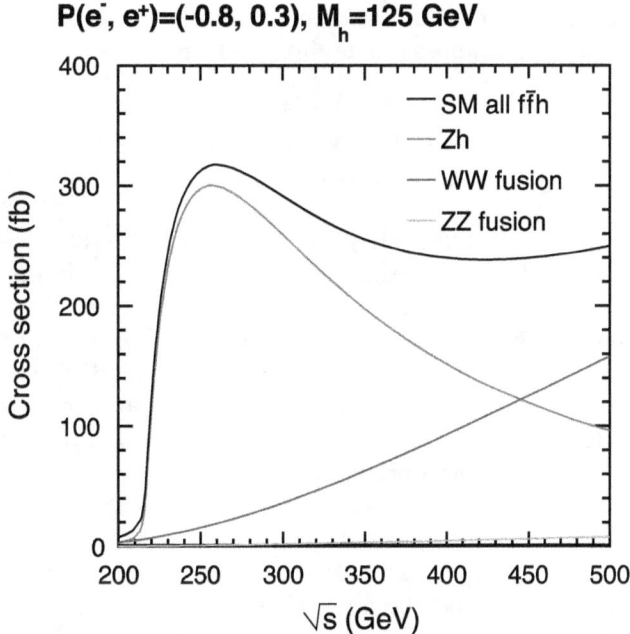

Figure 3.4. Cross sections of the processes $e^+e^- \rightarrow HZ$ (red curve), $e^+e^- \rightarrow H\nu_e\bar{\nu}_e$ (blue curve) and $e^+e^- \rightarrow He^+e^-$ (green curve) as a function of the center-of-mass energy for the beam polarizations $P(e^-, e^+) = (-0.8, +0.3)$. The total cross section is shown by a black curve. Source: ILC TDR, Volume 2, arXiv: 1306.6352, figure 2.7.

opposite values of P(e^-, e^+) = (+0.8, −0.3). The variation of polarization provides additional flexibility in many aspects. In particular, it is a very useful tool for coupling measurements. Potentially, specific SM or NP processes can be enhanced or suppressed by choosing the appropriate combination of the beam polarizations, which allows one to increase the signal and suppress backgrounds.

The Higgsstrahlung process comes via the s-channel so its cross section has a maximum just above the threshold, whereas the vector boson fusion is a t-channel process, which yields a cross section that grows logarithmically with the center-of-mass energy. As can be seen in figure 3.4, the ILC operating at 250 GeV will already have a high production rate of Higgs bosons. However, new channels of Higgs boson production will open up at higher energies, which will provide additional possibilities in the Higgs boson parameter measurements.

At 250 GeV, the backgrounds to the process $e^+e^- \rightarrow HZ$ are large and come mostly from the processes $e^+e^- \rightarrow W^+W^-$ and $e^+e^- \rightarrow Z\,Z$, which have large cross sections. The cross section of the W^+W^- process is approximately two orders of magnitude larger than that for the $e^+e^- \rightarrow HZ$ process; the ZZ production cross section is about 3–4 times larger. The backgrounds were simulated and experimental methods of their suppression in physical analyses were developed. Nevertheless, the residual contributions of these backgrounds are not small. Production of light fermion pairs also has large cross sections. However, these backgrounds can be effectively suppressed.

In case of an ILC upgrade to a higher energy, new channels with the production of the Higgs boson will open up and can be studied experimentally. The most sizeable channels and their approximate threshold values are:

(a) $E \approx 360\,\text{GeV}\ e^+e^- \rightarrow HHZ$

(b) $E \approx 460\,\text{GeV}\ e^+e^- \rightarrow HH\nu_e\bar{\nu}_e$ (3.11)

(c) $E \approx 480\,\text{GeV}\ e^+e^- \rightarrow Ht\,\bar{t}.$

The cross sections for these Higgs boson production channels, which can potentially be accessible for experimental observation, are shown in figure 3.5. As can be seen in the figure, the cross section for the process $e^+e^- \rightarrow Ht\bar{t}$ is about 2 fb at a maximum in the energy range (500–550) GeV. To observe this channel at least $\sim 100\,\text{fb}^{-1}$ of data taken with such an energy is required. The measurement of this cross section will allow us to directly measure the coupling g($Ht\bar{t}$), which will be an important test of BSM models. In addition, it is possible to search for a direct CP violation in the Higgs boson production in this channel. This will be discussed below. The cross section for the process $e^+e^- \rightarrow HHZ$ is typically of the order of 0.1 fb at a collision energy slightly above the threshold, which makes it difficult to observe this channel experimentally. However, it would be very important to observe a signal in this channel, because this will allow one to measure the Higgs boson self coupling. Potentially the triple Higgs boson coupling can also be obtained from a measurement of the $e^+e^- \rightarrow HH\nu_e\,\bar{\nu}_e$, but the cross section of this process is very small. Figure 3.6 shows the corresponding diagrams with a

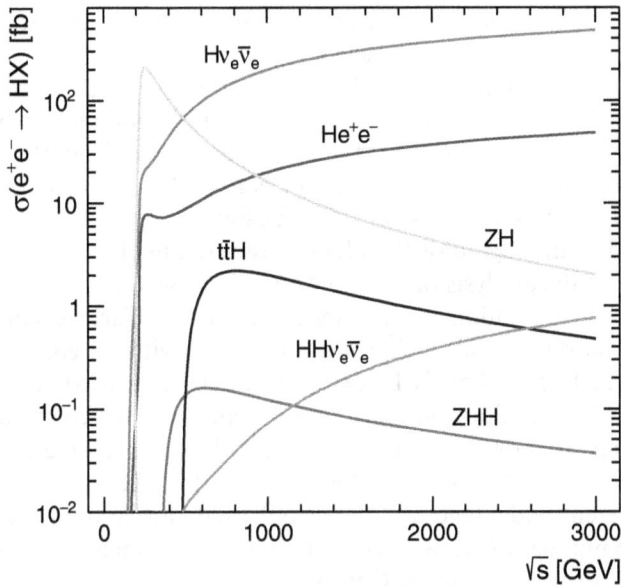

Figure 3.5. Cross sections for processes with the Higgs boson production in the e^+e^- collisions as a function of the center-of-mass energy. In the calculations, the unpolarized electron and positron beams were assumed. Source: CLIC programme, arXiv: 1209.2543, figure 2.1. © 2012–2018 CERN (License: CC-BY-3.0).

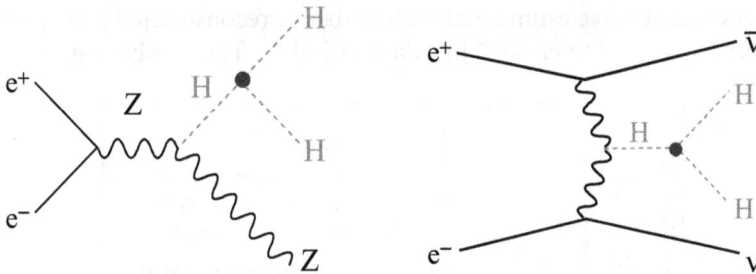

Figure 3.6. Diagrams describing processes with a triple Higgs vertex, $e^+e^- \to HHZ$ (left) and $e^+e^- \to HH\nu_e\bar{\nu}_e$ (right). The Higgs bosons are shown by red dotted lines, the blue dots indicate the triple Higgs boson couplings.

triple Higgs vertex, where the self coupling of the Higgs boson is marked by a blue dot.

3.1.4 Higgs boson decay branching fractions, mass and width

The precise experimental measurements of the Higgs boson couplings are the most important tasks at the ILC. In contrast to the LHC, the couplings can be measured at the ILC using a model-independent method. Moreover, the measurement accuracy at the ILC will exceed the accuracy that can be achieved at the LHC even after the planned luminosity upgrade. The experimental methods of the couplings measurements at different energies and, in particular at 250 GeV, were discussed in several ILC physics review papers.

The measurement of the $\mu^+\mu^-$ combination parameters is a key task in the Higgs boson studies at ILC. First, the mass of the $\mu^+\mu^-$ combination has to be measured, as well as the missing (also called recoil) mass to this system. The process $e^+e^- \rightarrow HZ$ with the following decay $Z \rightarrow \mu^+\mu^-$ has a clear signature: the $M(\mu^+\mu^-)$ mass peak will be close to the nominal mass of the Z boson and the recoil mass $M_{recoil}(\mu^+\mu^-)$ must be peaked in the region of the Higgs boson nominal mass.

The method of the analysis of the $\mu^+\mu^-$ combination was tested at the ILC using the Monte Carlo simulation. In the test, the Monte Carlo event samples were generated for the signal and background processes with an equivalent integrated luminosity of 250 fb^{-1}. The ILD detector model was used in simulation and reconstruction. The resulting $M_{recoil}(\mu^+\mu^-)$ distribution, which includes the signal and backgrounds, is shown in figure 3.7. The tail in the missing mass distribution corresponds to photon radiation in the initial state. The signal and background shapes are different and, therefore, the signal yield can be obtained from a fitting procedure. The number of signal events was extracted from a simultaneous fit to the mass and recoil mass distributions. Correspondingly, an efficiency of the $e^+e^- \rightarrow ZH$ process reconstruction was calculated. The signal and background shapes used in the fit were obtained from similar MC simulations.

Using this method, the uncertainty of the $e^+e^- \rightarrow ZH$ cross section measurement was projected to be equal to 1% for the total integrated luminosity of ~2 ab^{-1}. This value agrees with a naive estimate of the number of reconstructed $\mu^+\mu^-$ pairs, where this number is expected to be ~10^4 in a dataset 2 ab^{-1}. The number can be calculated

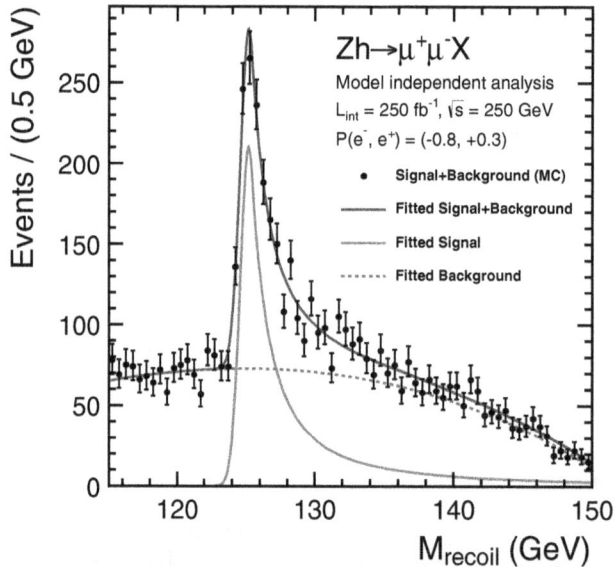

Figure 3.7. The distribution of the recoil mass to the system $\mu^+\mu^-$ is shown. The dataset for the signal process $e^+e^- \rightarrow ZH$ with the subsequent decay $Z \rightarrow \mu^+\mu^-$ was simulated assuming an energy of $\sqrt{s} = 250$ GeV and the integrated luminosity of 250 fb^{-1}. Source: K Fujii *et al*, arXiv: 1506.05992, figure 3.

as the product of the number of the Higgs bosons of 6×10^5 in the dataset (equation (3.2)), the $Z \rightarrow \mu^+\mu^-$ decay branching fraction $(3.366 \pm 0.007)\%$ and the reconstruction efficiency $\sim$50%. Correspondingly, the accuracy of the coupling g_Z measurement is also estimated as $1/\sqrt{10^4} \sim 1\%$.

The study of the $\mu^+\mu^-$ system allows us to measure several additional parameters of the Higgs boson. Firstly, the method provides a normalization to branching fraction measurements of other Higgs boson decay channels. Secondly, it is possible to obtain a decay branching fraction in the invisible final states. Thirdly, it allows us to measure the full width of the Higgs boson in a model-independent way. Today, the best experimental limit on the width of the Higgs boson given in the particle data group (PDG) is $\Gamma(H) < 22$ MeV at 95% CL, which is a combination of indirect measurements performed in the ATLAS and CMS experiments. The theory predicts a width of (4.1 ± 0.1) MeV. As estimated using the Monte Carlo simulation, the uncertainty of the width measurement of $\sim$13% can be obtained at the ILC with a dataset of 250 fb $^{-1}$ collected at 250 GeV. Fourth, the mass of the Higgs boson can be measured at the ILC with an uncertainty of $\sim$(30–40) MeV. Today, the combined mass value of the Higgs boson, including all results obtained by the ATLAS and CMS collaborations, is equal to $m(H) = 125.09 \pm 0.24(\pm 0.21 \pm 0.15)$ GeV (PDG). Therefore, the accuracy of the Higgs boson mass and width measurements at the ILC can be significantly improved compared with the LHC measurements.

The methods proposed to analyse the ILC data can be illustrated using table 3.2. The table includes parameters for several processes, which can be measured experimentally, and the theoretical expressions for these parameters, given as a function of the couplings and width of the Higgs boson.

In the table, g_Z, g_W and g_b are the Higgs boson couplings HZZ, HWW and $Hb\bar{b}$. Here Γ_H is the Higgs boson width and C_1, C_2, C_3 and C_4 are the constants that can be calculated theoretically with uncertainties of less than 1%.

Events corresponding to the processes given in table 3.2 can be selected experimentally and the number of signal events can be extracted using a fitting procedure for each process. From the numbers of the signal events, the experimental values for the products of cross sections and branching fractions can be obtained.

Table 3.2. The processes, the corresponding experimentally measurable parameters and their theoretical dependences on the couplings and width of the Higgs boson are presented.

	Process $e^+e^- \rightarrow$	Experiment, measured values	Expressed in couplings, Γ_H
(1)	$H(all) \, Z(\mu^+\mu^-)$	$\sigma(HZ) \cdot Br\,(Z \rightarrow \mu^+\mu^-)$	$= C_1 \cdot g_Z^2$
(2)	$H(b\bar{b}) \, Z(\mu^+\mu^-)$	$\sigma(HZ) \cdot Br(H \rightarrow b\bar{b}) \cdot Br\,(Z \rightarrow \mu^+\mu^-)$	$= C_2 \cdot g_Z^2 \, g_b^2 / \Gamma_H$
(3)	$H(b\bar{b}) \, \nu\bar{\nu}$	$\sigma(H\nu\bar{\nu}) \cdot Br\,(H \rightarrow b\bar{b})$	$= C_3 \cdot g_W^2 \, g_b^2 / \Gamma_H$
(4)	$H(WW^*) \, Z(\mu^+\mu^-)$	$\sigma(HZ) \cdot Br(H \rightarrow WW) \cdot Br\,(Z \rightarrow \mu^+\mu^-)$	$= C_4 \cdot g_Z^2 \, g_W^2 / \Gamma_H$
(5)	$H(\tau^+\tau^-) \, Z(all)$	$\sigma(HZ) \cdot Br(H \rightarrow \tau^+\tau^-)$	$= C_5 \cdot g_Z^2 \, g_\tau^2 / \Gamma_H$

Theoretically, these parameters can be expressed as functions of the corresponding couplings and the width of the Higgs boson, as shown in table 3.2. The constants C_i can be obtained from theoretical calculations because all other parameters are well known. It allows us to get the uncertainties in the calculations of these constants down to less than 1%. The squared coupling values enter into expressions from each Higgs boson vertex in the production and decay diagrams. The width Γ_H comes from the ratio:

$$Br(H \rightarrow X) = \Gamma_X/\Gamma_H, \qquad (3.12)$$

where Γ_X is the partial width, and $Br(H \rightarrow X)$ is the branching fraction of the decay $H \rightarrow X$.

The measurement of the process (1) from table 3.2 based on the $\mu^+\mu^-$ combination study was described above. The coupling g_Z will be obtained from this measurement with an accuracy of 1%. Then, because Γ_H is known theoretically with high accuracy, other couplings can be obtained from the measurements of the corresponding processes. For example, the couplings g_Z, g_W and g_τ can be obtained from the event ratios of the processes (2):(1), (4):(1) and (5):(1). It has to be noted, that large numbers of signal events are expected for all these processes, and the corresponding couplings can be measured in the ILC at 250 GeV with an accuracy close to the percentage level. Therefore, process (1) provided the normalization for the coupling measurements.

Using this method, the couplings and the branching fractions for different channels can be obtained. Summing up all measured branching fractions of the Higgs boson, the residual probability for decay in an invisible final state can be calculated. Again, the method uses the measurement of process (1) in the calculations.

To obtain the Higgs boson total width in a model-independent way, the ratio of processes (3):(2) has to be measured. This allows one to obtain the ratio g_W/g_Z (see table 3.2). The measurement of process (3) is methodically difficult because the background for this process is large, and reconstruction of two b-jets leads to large uncertainties. When the coupling g_Z is obtained from the measurement of process (1), the coupling g_W can be calculated from the ratio. For known couplings g_Z and g_W, the width Γ_H can be obtained from the measurement of process (4). The estimated accuracy of the width measurement is obtained as $\sim 13\%$ assuming a 250 fb^{-1} dataset, where the dominant uncertainty comes from the measurement of process (3).

Finally, the mass of the Higgs boson can be obtained directly from the $M_{\text{recoil}}(\mu^+\mu^-)$ distribution, assuming that the mass is a free parameter in the fit shown in figure 3.7. Taking into account an expected large data sample and low backgrounds, the mass of the Higgs boson can be precisely measured.

Potentially, the accuracy of the Higgs boson width measurement, comparable to the accuracy of the method discussed above, can be obtained from the measurement of the process $e^+e^- \rightarrow HZ$, with the following decay $H \rightarrow ZZ^*$, where the Z boson is reconstructed in the channels with hadronic jets. However, to obtain a good accuracy, the off-shell mass Z^* boson should be reconstructed in the lepton channels

Table 3.3. The branching fractions of the Z boson decays for the channels, which can be used to reconstruct the Higgs boson. These values are taken from the PDG.

$\mu^+\mu^-$	e^+e^-	$b\bar{b}$	$c\bar{c}$	hadrons
3.366 ± 0.007	3.363 ± 0.004	15.12 ± 0.05	12.03 ± 0.21	69.91 ± 0.06

$Z^* \to \mu^+\mu^-$ or $Z^* \to e^+e^-$. This method can provide sufficient background suppression and a good reconstruction efficiency. The product of the cross section and branching fraction for this process is proportional to g_Z^4/Γ_H. Then the width of the Higgs boson can be directly obtained using the coupling g_Z measured in process (1). The Z boson decay channels, which can be potentially used for this measurement, are given in table 3.3.

To visualize what the studied events look like in the ILD detector, the event display for the process $e^+e^- \to H(b\bar{b}) Z(\mu^+\mu^-)$ is shown in figure 3.8 for the center-of-mass energy 250 GeV (top) and 500 GeV (bottom). The events were simulated using the ILD detector model. The results of the reconstruction procedure are shown. The muons registered in the muon system are indicated. The b-jets are clearly visible as compressed angular correlated activites in the tracker, electromagnetic calorimeter, and hadron calorimeter.

3.1.5 Effective field theory formalism

The κ-formalism mentioned in subsection 3.1.1 is widely used in LHC measurements. If the modifiers κ_i deviate from unity, it indicates an NP contribution. Measuring the modifiers κ_i is one of the most important tasks at the ILC. However, there is a problem: the κ-formalism is based on the assumption that the same modifier can be applied in all processes, where the corresponding coupling is included. It works well as a first approximation, however κ-formalism is not a universal method to describe the couplings in all BSM models.

A more complicated approach of the couplings parametrization is proposed in the framework of Effective Field Theory (EFT). In this approach it is assumed that deviations from SM predictions can be described by anomalous contributions into the Lagrangian by adding the operators dimension-6. In this method the modifiers κ can still be used to describe the Higgs boson couplings to fermions. However, this is not enough to describe the Higgs boson couplings with the Z and W vector bosons. In particular, the interaction of the Higgs boson with the Z boson can be expressed in the EFT approach by a Lagrangian, including the term responsible for the anomalous tensor interaction:

$$\mathcal{L} = -\frac{m_Z^2}{v}(1 + \eta_Z) H Z_\mu Z^\mu + \frac{\zeta_Z}{2v} H Z_{\mu\nu}Z^{\mu\nu}. \tag{3.13}$$

Here m_Z is the mass of the Z boson, and η_Z and ζ_Z are the normalization constants, which correspond to the contributions of the standard and anomalous interactions,

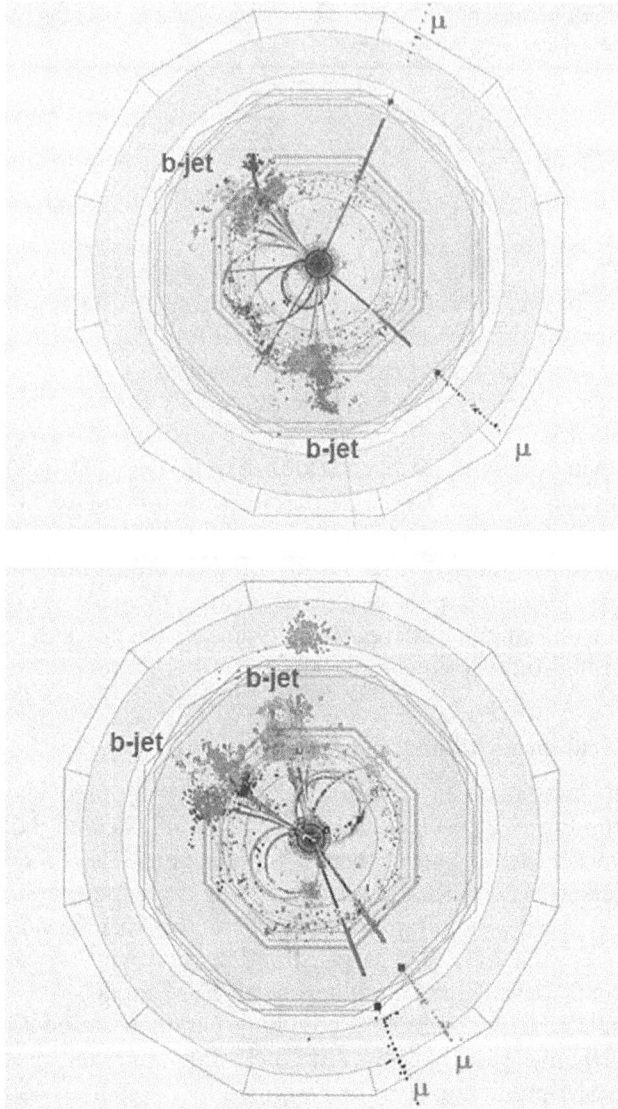

Figure 3.8. The event display for the process $e^+e^- \to HZ$ with the decay chain $H \to b\bar{b}$ and $Z \to \mu^+\mu^-$ at 250 GeV (top) and at 500 GeV (bottom) is shown. The detector responses due to muons and b-jets are clearly seen. The muons are marked red, the b-jets are marked blue.

respectively. Analogously, the interaction of the Higgs boson with the W boson can be expressed. In this approach, the parametrization of cross sections and widths has a more complicated form. In the EFT formalism, the coefficients before the normalization constants η_Z and ζ_Z can vary depending on the specific process and the collision energy. In particular, the Higgs boson production cross section and the partial decay width at 250 GeV will be parametrized differently:

$$\sigma(e^+e^- \to HZ) = (\text{SM}) \cdot (1 + 2\eta_Z + 5.7\zeta_Z)$$
$$\Gamma(H \to ZZ^*) = (\text{SM}) \cdot (1 + 2\eta_Z - 0.5\zeta_Z).$$

(3.14)

An adequate description of couplings in the framework of the EFT formalism requires a more complicated procedure to account correctly for all parameters.

3.1.6 Couplings measurements at LHC

In the ATLAS and CMS experiments at the LHC collider, signals with significances of more than 5σ have been observed in the decay channels $H \to \gamma\gamma$, $H \to ZZ^*$, $H \to WW^*$ and $H \to \tau^+\tau^-$. The channel $H \to b\bar{b}$ is seen with a signal significance larger than 3σ. In practice, the products of the cross section and branching fraction are usually measured in the LHC experiments. Therefore, it is convenient to obtain the signal strength, which is the ratio of the product measured experimentally to the product predicted in the framework of the SM:

$$\mu = (\sigma \cdot Br)_{\text{meas}}/(\sigma \cdot Br)_{\text{SM}}.$$

(3.15)

If the measured value agrees with the SM prediction, the corresponding signal strength is equal to unity, $\mu_i = 1$. The μ value, averaged over all channels observed and including all results obtained in both LHC experiments, is equal to $\mu = 1.09 \pm 0.07 \pm 0.09$. This value demonstrates a good agreement of experimental data with SM predictions.

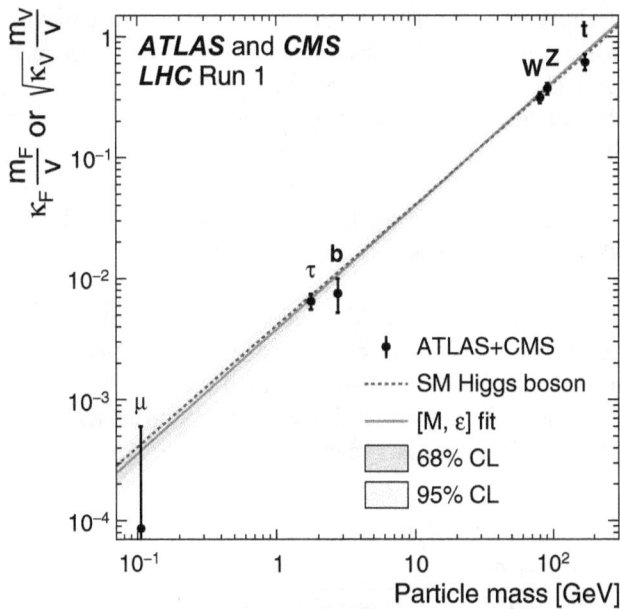

Figure 3.9. The combination of ATLAS and CMS results of the Higgs couplings measurements is shown. In the SM, the Higgs boson couplings are proportional to fermion masses and squared boson masses. The shown data confirm this dependence. Source: JHEP08 (2016) 045, arXiv: 1606.02266, figure 19. A combined ATLAS and CMS analysis of *pp* collision data at 7 and 8 TeV © 2018 CERN.

If the cross section is known from theoretical calculations or other measurements, the branching fraction can be obtained from the signal strength for each channel. Then the corresponding Higgs couplings can be recalculated from the branching fractions. Recent ATLAS and CMS measurements of the couplings are shown as a function of the final state masses in figure 3.9. Within large experimental uncertainties, a good agreement with the linear dependence is seen. The modifiers κ_i were also calculated using the LHC coupling measurements. Their values coincide with unity within the experimental uncertainties.

3.1.7 Why ILC construction is important

Using Monte Carlo simulation, the potential accuracy of the Higgs boson couplings measurements at the ILC was estimated within the EFT formalism. To obtain the uncertainties, a data sample equivalent to 2 ab^{-1} was generated, where the collision energy 250 GeV and polarization P(e^-, e^+) = (-0.8, $+0.3$) were assumed. Table 3.4 shows the expected uncertainties of the coupling measurements at the ILC for different Higgs boson decay channels.

As can be seen from the table, most of the experimental uncertainties are at a level of 1%. It is very important to reach the 1% level in the accuracy of the couplings measurements. A comparison of the experimental uncertainties, which can be achieved at LHC after upgrade and a combination of the LHC and ILC measurements, is shown in figure 3.10. As can be seen from the figure, a significant improvement in the accuracy of the couplings measurements can be achieved. Using only the LHC data, the uncertainties will be about 4%, whereas the combination of the LHC and ILC data will allow us to achieve an accuracy of 1%. It is worth recalling that the deviations of the Higgs boson couplings in the different BSM models (see figure 3.2) are of the order 5%. Therefore, only the LHC data will not be sensitive to coupling deviations expected in the BSM models. However, the inclusion of the ILC data will provide a crucial test for many BSM models. It is possible that the projections of the LHC uncertainties were done conservatively, and the LHC experiments will measure with full statistics the Higgs boson couplings with an accuracy of ~(2–3)%, however this does not change the final conclusion.

It is worth repeating that all BSM models discussed above must contain new heavy Higgs bosons with a mass below 5 TeV or maybe, in some additional assumptions, below 10 TeV. With an increasing mass limit for the new heavy Higgs bosons, the deviations of the couplings from SM predictions will become smaller and probably disappear when the masses reach values of about 5 TeV. As was mentioned

Table 3.4. The expected uncertainties (in %) in the Higgs boson couplings measurements at ILC, obtained in the framework of the EFT approach. Source: Barklow T *et al* 2018 *Phys. Rev.* D **97** 053003, Table 3.

g($Hb\bar{b}$)	g($Hc\bar{c}$)	g(Hgg)	g(HWW^*)	g($H\tau\tau$)	g(HZZ)	g($H\gamma\gamma$)	g($H\mu\mu$)
1.04	1.79	1.60	0.65	1.16	0.66	1.20	5.53

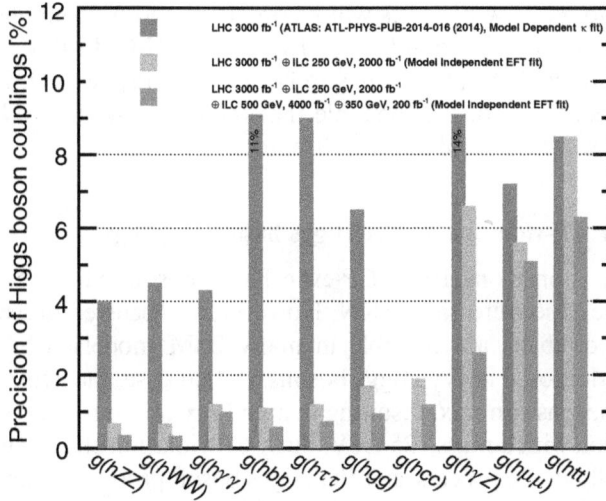

Figure 3.10. The accuracy of the determination of the couplings of the Higgs boson obtained in the framework of the EFT formalism is shown. The LHC uncertainties were obtained in the ATLAS collaboration assuming the HL-LHC luminosity upgrade and full statistics of 3000 fb^{-1}. These uncertainties are compared with the uncertainties expected for a combination of measurements from the LHC and ILC, in particular, assuming that the statistics of 2 ab^{-1} will be collected in ILC at 250 GeV. Source: K Fujii *et al*, arXiv: 1710.07621, figure 5.

above, by now the LHC experiments do not see any NP particles up to a level of about 1 TeV. Although of course, each specific upper limit depends on the decay channel studied. Over the next decade, the upper limits of the LHC's searches for BSM particles will come up to a level of about 2 TeV. However, the potential LHC limits will still be below the theoretical limits for some BSM models.

There are many experiments, which have sensitivity to NP effects. In particular, in the fields of *B* physics, neutrino studies, lepton flavor violation searches and some other areas. Certainly all these experiments will critically test different scenarios of the BSM physics models. However, only two experiments have the highest sensitivity to BSM models with the extended Higgs sector: the LHC and potentially the ILC. The LHC experiments are most sensitive to the direct production of heavy Higgs bosons, and the ILC is the most sensitive in indirect searches of heavy Higgs bosons by precisely measuring the 125 GeV Higgs boson couplings.

If the ILC is constructed, there are three possible scenarios in the BSM physics searches. In the first scenario, the LHC will find at least one new Higgs boson with a mass in the range of (1–2) TeV. Then the couplings and other parameter measurements in the ILC will allow us to choose the correct BSM model. In the second scenario, the LHC will not see new Higgs bosons, but the coupling deviations will be found at the ILC and this will allow us to find a most appropriate BSM model without discovering new Higgs bosons. In both cases, the answer will be given to a very important question: if the Higgs boson is a point-like or composite particle. In the third scenario, no new particles will be observed and no coupling deviations or any other NP effects will be found at the LHC or the ILC. In this case, the current theoretical picture of BSM physics has to be substantially revised. In the case that the ILC is not constructed, it is possible that the question about NP and theories beyond the SM will be postponed for 30 years.

3.1.8 Search for *CP*-violation in SM Higgs boson couplings

In the SM there is only one 100% *CP*-even Higgs boson, which is identical to the 125 GeV particle, discovered at CERN. However, as discussed in section 3.1.2, two or more Higgs doublets are possible in many BSM models, which leads to the appearance of additional heavy Higgs bosons. In that case, the neutral *CP*-even and *CP*-odd Higgs bosons can mix, resulting in an admixture of the *CP*-odd component in the Higgs boson with a mass 125 GeV. Such *CP* violation effects should be looked for experimentally in the processes of the Higgs boson production and decays. Were this *CP*-odd admixture to be found, this would be a direct manifestation of NP. If only a strict upper limit is set on this admixture, it will constrain BSM models and, probably, some of these models will simply be excluded.

The *CP*-odd component exhibits itself in the Higgs boson couplings or, in other words, in processes with the Higgs boson production or decay. The angular distributions of the initial and final state particles in the processes with the Higgs boson must be studied to search for the *CP*-violation effects. There are some other characteristics, which can also be used for such searches, especially in the case of interference of several amplitudes (diagrams). For example, the cross section of the process $e^+e^- \to HZ$ can grow differently near the production threshold for different *CP*. However, the number of parameters and processes, which are really sensitive to the *CP*-odd component contribution, are limited.

The spin information is often lost during hadronization, when the final state particles have zero spins. In some cases, it is not possible to measure the spin configuration. However, the *CP*-odd component can exhibit itself in angular distributions in the processes, which include the $H\tau\tau$, $Ht\bar{t}$, HZZ and HWW couplings. The couplings HZZ and HWW are not well suited for a precise measurement of the fraction of the *CP*-odd component, because these contributions are strongly suppressed. The coupling $H\tau\tau$ is most convenient for such studies, since the decay $H \to \tau^+\tau^-$ has a large branching fraction and τ decay products keep information about the τ spin direction. Information about *CP* composition can also be extracted from processes containing the $Ht\bar{t}$ coupling, because the top quark decays before hadronization and its decay products retain information about the top quark spin. Unfortunately, the Higgs boson production processes, where the $Ht\bar{t}$ coupling appears, have small cross sections.

The most appropriate processes that can be used at ILC for *CP*-violation searches are:

- Decay $H \to ZZ^*$ with following $Z / Z^* \to \ell^+\ell^-$ decays, where $\ell^\pm = e^\pm / \mu^\pm$.
- Process $e^+e^- \to HZ$ with following $Z \to \ell^+\ell^-$ decay, where $\ell^\pm = e^\pm / \mu^\pm$.

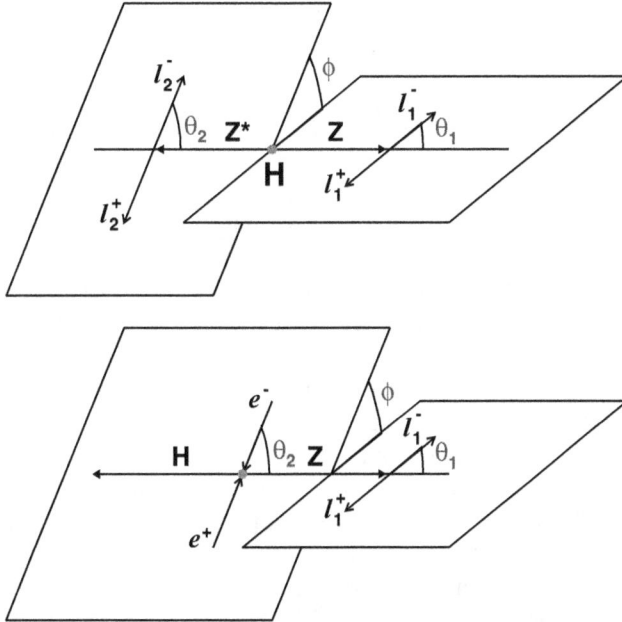

Figure 3.11. The angles are shown, which can be used to measure the CP violation in the $H \rightarrow ZZ^*$ decay (top) and the process $e^+e^- \rightarrow HZ(\ell\ell)$ (bottom): the helicity angles θ_1 and θ_2 and the angle between the decay planes ϕ. The coupling HZZ is indicated by red dots.

- Decay $H \rightarrow \tau^+\tau^-$ with following $\tau^\pm \rightarrow a^\pm\nu_\tau$, where $a^\pm = \pi^\pm, \rho^\pm, a_1^\pm, \ell^\pm\nu_\ell$.
- Process $e^+e^- \rightarrow Ht\bar{t}$.

The $H \rightarrow ZZ^*$ decay into four leptons in the final state is the simplest process for such analysis because all the particles are well measured. A standard set of the helicity angles and the angle between the decay planes can be used (see figure 3.11 (top)) to test the CP content in the HZZ coupling. Generally speaking, other variables can also be tested to search for the CP-odd component of the Higgs boson, however, the interpretation of their distributions will require comparison with model-dependent calculations.

The situation with the decay $H \rightarrow ZZ^*$ is quite ambiguous. It is difficult to obtain a quantitative estimate of the CP-odd contribution in this decay. The CP-even contribution in the HZZ coupling is described by a tree diagram, whereas the CP-odd component can contribute only via a strongly suppressed loop diagram. As a result, the contribution of the CP-odd component is initially strongly suppressed, and therefore it is difficult to quantitatively evaluate the CP-odd component using the $H \rightarrow ZZ^*$ decay.

On the other hand, the CP-odd component in the decay of $H \rightarrow ZZ^*$ is allowed in the EFT approach. In this approach, the CP-odd contribution can be several percent. In the EFT approach, the Lagrangian used to describe this decay can include an anomalous contribution, which is represented by two terms corresponding to the *CP*-conserving and *CP*-violating tensor components:

$$\mathcal{L} = -\frac{m_Z^2}{v}\left(1 + a_Z\right) H Z_\mu Z^\mu + \frac{b_Z}{2v} H Z_{\mu\nu} Z^{\mu\nu} + \frac{\tilde{b}_Z}{2v} H Z_{\mu\nu} \tilde{Z}^{\mu\nu} \qquad (3.16)$$

where a_Z is a renormalization factor, b_Z corresponds to the *CP*-even tensor component, and $\tilde{b}_Z$ is responsible for the *CP*-odd contribution.

The helicity angles θ_1 and θ_2 and the angle between the decay planes ϕ have different distributions for the three terms given in equation (3.16), as shown in figure 3.12. Therefore, the *CP*-odd component can be extracted from the measurements of these angular distributions.

In the ILC, the number of reconstructed Higgs bosons in the decay $H \rightarrow ZZ^*$ will be small. This decay is better suited for analysis at the LHC. This decay was analysed in the ATLAS and CMS experiments. As explained above, the fraction of the *CP*-odd component cannot be obtained in this measurement. However, it can be proven that the 125 GeV Higgs boson is not a pure *CP*-odd state with the quantum numbers $J^p = 0^-$. In the measurements performed in the ATLAS and CMC experiments, confidence levels of 97.8% *CL* and 97.6% *CL*$_s$ were obtained. These values reflect the experimentally obtained high probability that the Higgs boson is not a pure *CP*-odd state.

At the ILC, the coupling *HZZ* can be tested in the process $e^+e^- \rightarrow ZH$, with subsequent decays $Z \rightarrow \mu^+\mu^-$ or $Z \rightarrow e^+e^-$. As in the previous case, the same problem appears in this measurement. Again, the *CP*-odd contribution is suppressed because of the loop diagram and its admixture cannot be well defined. Nevertheless, within the framework of the EFT approach, the *CP*-odd component can give an

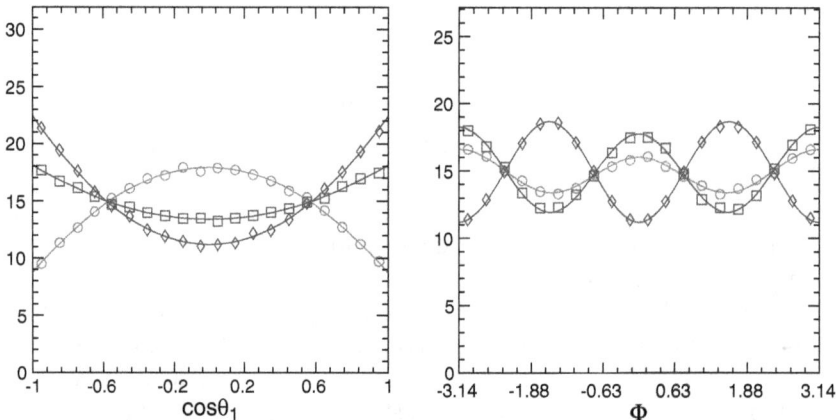

Figure 3.12. The predicted angular distributions for the helicity angles θ_1 and θ_2 (left) and the angle between the decay planes ϕ (right). The distributions are shown for the SM Higgs boson (red circles), for the *CP*-conserving anomalous contribution corresponding to the second term in equation (3.16) (green squares), and *CP*-violating anomalous contribution corresponding to the third term in equation (3.16) (blue diamonds). Reprinted with permission from S Volognesi *et al* 2018 *Phys. Rev.* D **86** 095031, figure 12 (the leftmost bottom figure and left next from bottom). Copyright 2012 by the American Physical Society.

observable contribution. At the ILC, a large number of events will be reconstructed in the process $e^+e^- \rightarrow ZH$. Accordingly, a high accuracy in the CP-odd component measurement can be achieved. The angles sensitive to the CP-odd component are shown in figure 3.11 (bottom) and the predicted angular distributions are similar to those shown in figure 3.12.

At the ILC the most effective way to measure the CP violating component in the Higgs boson is to study the decay $H \rightarrow \tau^+\tau^-$. This decay has a relatively large branching fraction, and the τ decay products allow us to measure the τ lepton spin. For the angular analysis, the decays $\tau^\pm \rightarrow a^\pm \nu_\tau$ can be used, where $a^\pm = \pi^\pm, \rho^\pm, a_1^\pm, \ell^\pm \nu_\ell$. Potentially, all these channels have to be tested and combined into a common analysis. However, each channel has specific problems.

The Lagrangian for the decay $H \rightarrow \tau^+\tau^-$ can be expressed as:

$$\mathcal{L} = -\frac{m_\tau}{v} H \, \bar{\tau} \, (\cos \psi_{CP} + i \sin \psi_{CP} \, \gamma^5) \, \tau \tag{3.17}$$

where m_τ is the mass of the τ lepton, and the angle ψ_{CP} characterizes the fraction of the CP-odd admixture in the Higgs boson. If the mixing angle is $\psi_{CP} = 0$, the Higgs boson is a pure CP-even state. If the angle is $\psi_{CP} = \pi/2$, then it is a pure CP-odd state.

The main problem with this analysis is that the neutrinos from the τ decays are not registered. However, the τ decay vertex is displaced from the Higgs boson decay vertex by a few millimeters on average, which allows us to reconstruct the decay planes with a reasonably good accuracy. In the case of the decay $\tau^\pm \rightarrow \pi^\pm \nu_\tau$, the decay plane is fixed using the coordinates of the τ production vertex and the position and direction of the charged pion momentum vector.

In the process $e^+e^- \rightarrow HZ$, with a well measured primary vertex and the decay chain $H \rightarrow \tau^+\tau^-$, $\tau^\pm \rightarrow \pi^\pm \nu_\tau$, all parameters of the particles can be calculated using the reconstructed decay planes and conservation laws. The potentially measured angles are shown in figure 3.13. In the case of the $H \rightarrow \tau^+\tau^-$ analysis, the angle ϕ between the decay planes is the most sensitive for the CP composition measurement.

In this method we can calculate the angle ϕ and measure the possible contribution of the CP-odd component. The distribution for the angle ϕ^*, which is equal to the angle ϕ up to a shift to constant 2π, is described by the formula:

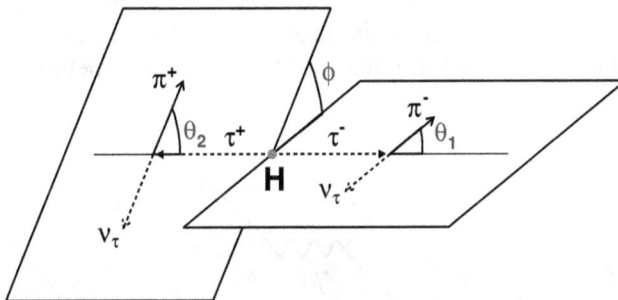

Figure 3.13. The angles measured in the $H \rightarrow \tau^+\tau^-$ decay: the helicity angles θ_1 and θ_2 and the angle between the decay planes ϕ. The coupling $H\tau\tau$ is shown by a red dot.

$$dN/d\phi^* = A + B \cdot \cos(\phi^* - 2\,\psi_{CP}) \qquad (3.18)$$

where the coefficients A and B are constants, with the approximate ratio of $A/B \approx 1.5$. The accuracy of the ψ_{CP} parameter measurement was tested by the Monte Carlo simulation method using the ILD detector model and a data sample with an integrated luminosity 2 ab^{-1} at an energy of 250 GeV. It was obtained that an accuracy of 4.3° can be achieved for the mixing angle ψ_{CP}.

It should be noted that interference of the CP-even and CP-odd amplitudes will result in asymmetric angular distributions, whereas pure CP states give angular distributions with a symmetric form. This allows us to construct model-independent parameters, which deviate from zero in the case where the 125 GeV Higgs boson contains both components. In all three processes discussed above, where angular distributions are studied, there are background contributions. The backgrounds are mostly combinatorial (resulting from a random combination of uncorrelated particles) and, therefore, flatly distributed. However, the shapes and magnitudes of all potential backgrounds must be accurately evaluated for analyses of angular distributions.

The process $e^+e^- \rightarrow Ht\bar{t}$ can also be used to test the CP-odd contribution in the Higgs boson coupling $Ht\bar{t}$. However, an energy of at least 500 GeV is required to measure this process. In this process, it is possible to search for a direct CP-violation, since the three diagrams contribute to this process (figure 3.14).

The last diagram includes the ZZH vertex, where, as discussed earlier, the CP-odd contribution is suppressed due to the loop diagram. At the same time, the first two diagrams include the $Ht\bar{t}$ vertex, where the CP-odd contribution is described by an unsuppressed tree diagram. The interference of these three diagrams can lead to a direct CP violation. To observe the CP-violation, it is necessary to measure the normalized triple product:

$$O = \frac{\vec{p}_t \cdot (\vec{p}_{e^-} \times \vec{p}_{\bar{t}})}{|\vec{p}_t| \cdot |\vec{p}_{e^-}| \cdot |\vec{p}_{\bar{t}}|}, \qquad (3.19)$$

where $\vec{p}_t$, $\vec{p}_{\bar{t}}$ and $\vec{p}_{e^-}$ are the momenta of the top quark, the anti-top quark and the beam electron, respectively. The normalized triple product, independently of the model, must be zero if CP is conserved and will differ from zero in the case of CP-violation. Generally speaking, there are other ways of searching for a CP-violation in the process $e^+e^- \rightarrow Ht\bar{t}$. In particular, this can be done by measuring the energy dependence of the cross section of the process and the angular asymmetry in the production of the top and anti-top quarks. However, these methods require precise

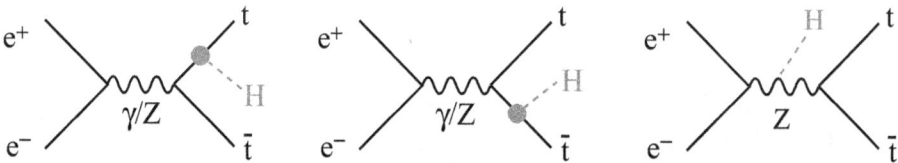

Figure 3.14. Diagrams describing the process $e^+e^- \rightarrow t\bar{t}H$. The Higgs boson is shown in red.

theoretical predictions for these parameters in the case of *CP*-odd contribution and, thus, these methods are model-dependent.

It should be noted that *CP*-violation can also be searched for in the process of the top quark pair production $e^+e^- \rightarrow t\bar{t}$. In this process, the *CP* violation can be induced by a loop diagram with a *CP*-violating vertex. The contribution of such diagrams is expected to be small in comparison with the regular tree diagram and, accordingly, the *CP*-odd component will be small. However, despite the small *CP* violation in this process, it will be partially compensated by the large cross section of the process.

3.1.9 Search for Higgs boson rare decays

The ILC is often called the 'Higgs factory', because a large number of the produced Higgs bosons is expected. If statistics of the order 2 ab^{-1} are obtained at 250 GeV, then the collected data sample will contain approximately 6×10^5 Higgs bosons. Therefore, rare decays with branching fractions down to $(2-20) \times 10^{-5}$ can be searched for, depending on the signal reconstruction efficiency and the amount of background contributions.

One of the first rare decays, which can be searched for in the ILC, is the decay $H \rightarrow Z\gamma$, which is predicted to have a branching fraction of $\sim 1.5 \times 10^{-3}$. Unfortunately, the clean *Z* decay modes, $Z \rightarrow e^+e^-/\mu^+\mu^-$ have low branching fractions of $\sim 3.7\%$ for each channel, which will result in a small number of reconstructed events. With a data sample of ~ 2 ab^{-1}, only ~ 40 events are expected, and these are difficult to observe. To increase the number of events, it is necessary to reconstruct the hadronic *Z* decay modes, in particular the two-jet $Z \rightarrow b\bar{b}/c\bar{c}$ modes with the branching fractions of $\sim 15\%$ and $\sim 12\%$, respectively. However, significant backgrounds are expected in the hadronic *Z* decay modes, which must be effectively suppressed. A direct photon can be imitated by an initial state radiative (ISR) photon. This effect also gives a significant background contribution. The decay channel $H \rightarrow Z\gamma$ can be observed with the full ILC statistics, however, it is a methodically difficult task.

The decay $H \rightarrow \mu^+\mu^-$ is of special interest because it allows us to measure the $H\mu\mu$ coupling. The theoretically expected branching fraction for this process is 2.17×10^{-4}. With full ILC statistics, the expected number of reconstructed events will be slightly less that 100. However, the background contributions are small. The main background comes from the decays $Z \rightarrow \mu^+\mu^-$ and $W^\pm \rightarrow \mu^\pm\nu$, where the energetic muons are produced. Potentially the decay $H \rightarrow \mu^+\mu^-$ should be observed at the ILC, taking into account the excellent momentum resolution in the ILD tracker.

The decay $H \rightarrow e^+e^-$ is not expected to be observed at the ILC, since its branching fraction is almost 40 000 times less than that for the $H \rightarrow \mu^+\mu^-$ mode. However, the $H \rightarrow e^+e^-$ decay has to be searched for, taking into account the recent discussion of a violation of lepton universality. The lepton universality assumes that all behaviours of the three leptons, the electron, the muon and the τ lepton, are the same with the difference arising only from the masses. Recently an evidence of the lepton non-universality has been found in the *B* meson decays.

However, the statistical significance of the results is not large enough to declare the observation. Therefore, it is important to test all processes with leptons in final states, especially at high energies.

Another Higgs boson decay, $H \rightarrow J/\psi\gamma$, cannot be seen at the ILC either, assuming the SM branching fraction is of ~2.5×10^{-6}. However, as well as for the previous decay modes, this process can be enhanced in specific BSM models. Therefore, the decay has to be searched for with the full ILC data sample.

In addition to the rare Higgs boson decays, a number of exotic modes, forbidden or strongly suppressed in the SM, can be searched for. There are no definite predictions for such processes, the expected branching fractions depend on the specific BSM model. The first such decay, $H \rightarrow \tau^{\pm}\mu^{\mp}$, has different leptons in the final state, which is treated as the lepton flavor violation (LVF). Although it has a negligibly small branching fraction in the SM, the LVF processes can be strongly enhanced in many BSM models, especially for the heavy leptons, just as in this decay. The decay $H \rightarrow b\bar{s}$ is also strongly suppressed in the SM with a branching fraction of about 10^{-7}, which can be enhanced in BSM models. This process is induced by a flavor changing neutral current (FCNC), for which a tree diagram is forbidden in the SM, but it is allowed in some BSM models. The FCNC processes are possible in specific BSM models for decays of the heavy Higgs boson H_2 and the heavy Z', which can be somehow reflected in the 125 GeV Higgs boson. It is possible to search for the Higgs boson decays due to LFV or FCNC effects in the case of light leptons or quarks. However, such decays are expected to be even more strongly suppressed.

3.2 Top quark studies

The top quark is the heaviest experimentally observed fundamental particle, this generates a special interest in this object. An accurate measurement of top quark parameters is extremely important theoretically. In particular, the mass of the top quark is a fundamental parameter of the electroweak theory. Amplitudes of loop diagrams depend on the masses of the particles in the loop. Because of the large mass of the top quark, the loop diagram amplitude with the top quark usually dominates over the other diagram amplitudes. Therefore, the mass value of the top quark essentially affects the results of calculations for many processes, which include loop corrections.

3.2.1 Top quark production and decay processes

At the first stage of the ILC operation, the collision energy is proposed to be limited to 250 GeV, and the top quark studies cannot be performed. Therefore, only a brief review of the most important studies of the top quark will be discussed here, which can be performed in the next stage when the collision energy will reach at least 380 GeV. The cross section of the top pair production process, $e^+e^- \rightarrow t\bar{t}$, will have a maximum of about 700 fb just after the threshold, at approximately 380 GeV (figure 3.15). Even with the collected integrated luminosity ~200 fb^{-1}, this data sample will contain ~2.8×10^5 of top quarks. At higher energies, two other top quark

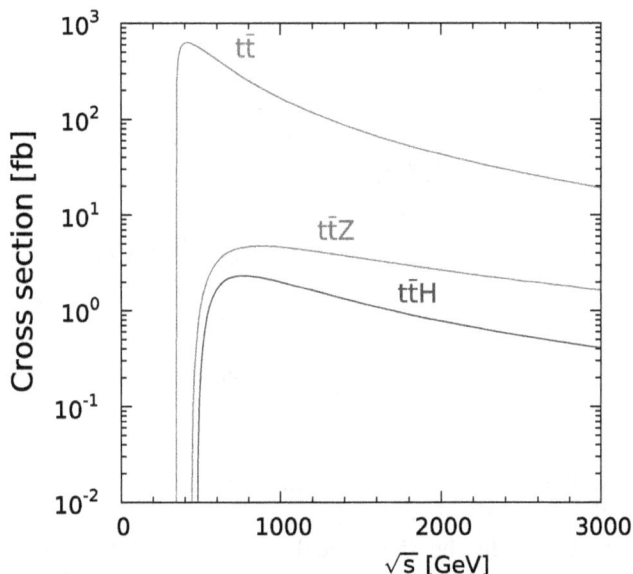

Figure 3.15. Cross sections of processes with the production of the top quarks, as a function of the e^+e^- collision energy. Curves are obtained from the MC simulation using the Whizard 2 event generator. Source: V Vos *et al*, arXiv: 1604.08122, figure 4.

production channels will open up: $e^+e^- \to t\bar{t}H$ and $e^+e^- \to t\bar{t}Z$, however, their cross sections are much smaller (figure 3.15). Nevertheless, these processes are of great interest, because their measurements can provide important information about the properties of the top quark. In particular, the process $e^+e^- \to t\bar{t}H$ allows one to perform a direct measurement of the $Ht\bar{t}$ coupling. In the TDR it was planned to take data at 500 GeV to measure this process. However, later studies showed that the twice better accuracy of the $Ht\bar{t}$ coupling measurement can be obtained at 550 GeV with the same amount of integrated luminosity.

The dominant top quark decay channel is $t \to Wb$, where the W boson can be reconstructed in semileptonic modes with unregistered neutrinos $W^\pm \to \mu^\pm \nu_\mu$ and $W^\pm \to e^\pm \nu_e$, or in hadronic modes, mostly in $W^+ \to c\bar{s}$. Although backgrounds for these processes can be significantly suppressed, the accuracy of W parameters reconstruction in both cases has large uncertainties. A diagram of the top quark production and decays into the semileptonic channel in e^+e^- collisions is shown in figure 3.16.

3.2.2 Top mass and width measurements

Interpretation of the mass of the top quark is very ambiguous. There are a number of theoretical definitions of the mass of the top quark. In particular, the mass of the top quark can be interpreted as the value of the mass pole, but this definition has significant uncertainties of the order Λ_{QCD}. Alternatively, the mass of the top quark can be determined from the peak position of the 1S-resonance mass, or the mass can be defined as the mass $\overline{\text{MS}}$, as determined in the renormalization scheme. The

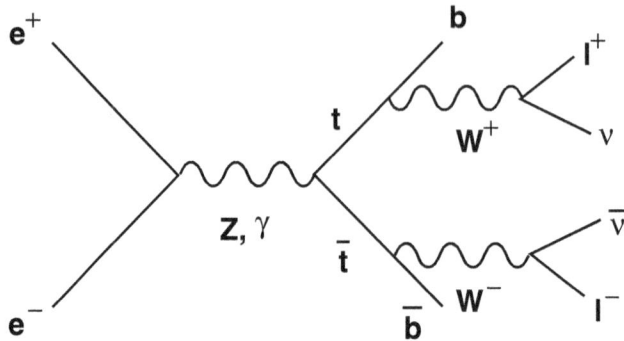

Figure 3.16. A diagram describing the top quark pair production and semileptonic decays. Reprinted from Atwood *et al* 2001 *Phys. Rep.* **347** 1–2, with permission from Elsevier.

difference between the last two top quark mass values was theoretically studied and found to be small, within 10 MeV.

It is important to relate the theoretical definitions of the top quark mass to experimental measurements. At hadronic colliders, the mass of the top quark can be measured by reconstructing the decay products. However, this method gives the mass value, which is set in the Monte Carlo simulation, and it is difficult to relate this experimental value to the theoretical definitions of the mass of the top quark. The discrepancy between this experimental value and the theoretical definitions can reach 1 GeV. The systematic uncertainties in the experimental measurements of the top quark mass at hadronic colliders are now slightly less than 0.5 GeV, and in prospects will be improved to 0.2 GeV in the future after the high luminosity LHC upgrade.

Another method of the top quark mass measurement at hadron colliders is to compare the $t\bar{t}$ pair production cross section with theoretical calculations, which include the mass of the top quark. This method also has large uncertainties and is model-dependent.

Much more accurately, and with a better relation to theoretical definitions, the mass of the top quark can be measured at e^+e^- colliders. For this, the $t\bar{t}$ pair production cross section has to be measured at several points of the collision energy near the mass threshold (figure 3.17). This method allows us to obtain a perfect experimental accuracy of the top mass measurement. The uncertainty of ~20 MeV in the top mass value can be achieved by measuring 10 points near the threshold with an integrated luminosity of ~10 fb^{-1} at each point. Figure 3.17 shows a simulation of such a comparison of experimental data with theoretical predictions. Moreover, this method allows us to obtain the experimental value of the top mass, which can be matched to the theoretical value defined within the 1S-resonance top mass interpretation. The total theoretical uncertainty, including the accuracy of the calculations performed and the uncertainty due to a difference between the experimental and theoretical values, in this method is ~50 MeV.

An important factor for using this method is the high accuracy of the theoretical predictions. Recently, significant progress has been achieved in this direction. The

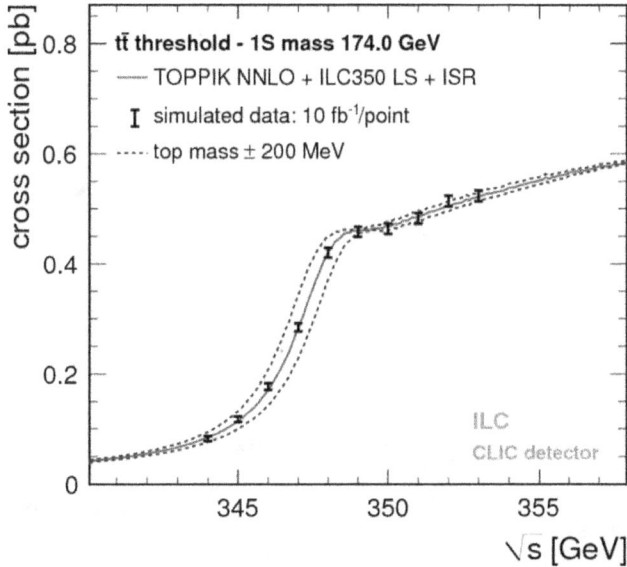

Figure 3.17. Simulation of the experimental measurement of the cross section of the $t\bar{t}$ pair production in the e^+e^- collisions. The mass of the top quark in the simulation is 174 GeV. The cross section is measured at 10 points near the threshold with an integrated luminosity ~10 fb^{-1} at each point. The values are compared to the theoretical predictions obtained for the top quark mass of 174 GeV and ± 200 MeV from the central value. Source: K Seidel 2013 *Eur. Phys. J.* **C73** 2530, figure 5.

top pair production cross section was calculated within the framework of the NLO, NNLO, and N^{3}LO approximations. A significant uncertainty in these calculations comes from the uncertainty in the α_s measurements, however, the accuracy of this parameter will be significantly improved over the next decade before the time when the ILC begins operation.

The width of the top quark is also an important parameter of the theory. Direct measurement of the width at hadron colliders is methodically difficult and leads to large uncertainties. The width can be obtained indirectly by measuring the cross section of a single top quark production. The average PDG value, $\Gamma_t = 1.41^{+0.19}_{-0.15}$ GeV/c^2, includes all measurements performed at hadron colliders using both direct and indirect methods. This value agrees with theoretical predictions within the large uncertainties. Again, a better accuracy of the top width measurement can be obtained in the e^+e^- collider ILC. The evolution of the theoretical predictions for different top quark width values is shown in figure 3.18. With increasing width, the growth of the cross section in the region of the 1S resonance becomes less steep. The width can be also extracted from a fit to the top pair production cross section near the threshold, similar to the top mass measurement.

It should be noted that the contribution of the loop diagram with the Higgs boson exchange between the top and anti-top quarks is of the order of 10% relative to the tree amplitude. The contribution leads to an increase in the $t\bar{t}$ production cross section, both at high energies and near the threshold (figure 3.19).

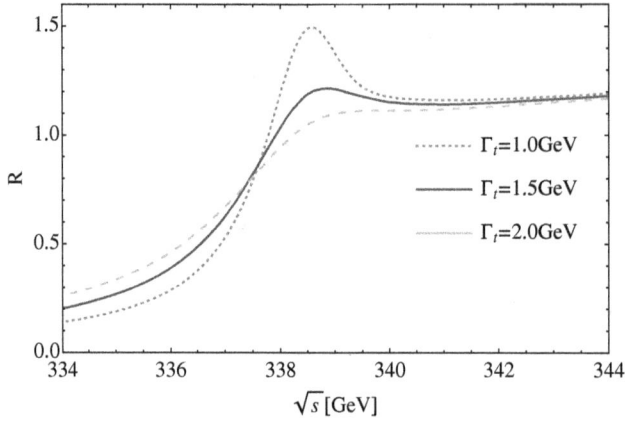

Figure 3.18. Calculations of the cross section for the $t\bar{t}$ pair production, normalized to the value $\sigma(\mu^+\mu^-)_{\text{QED}}$, as a function of the e^+e^- collision energy for different values of the width of the top quark. In the simulation, the values $m_t = 170$ GeV and α_s (30 GeV) = 0.142 were assumed. Source: G Moortgat-Pick *et al* 2015 *Eur. Phys. J.* **C75** 371, figure 101.

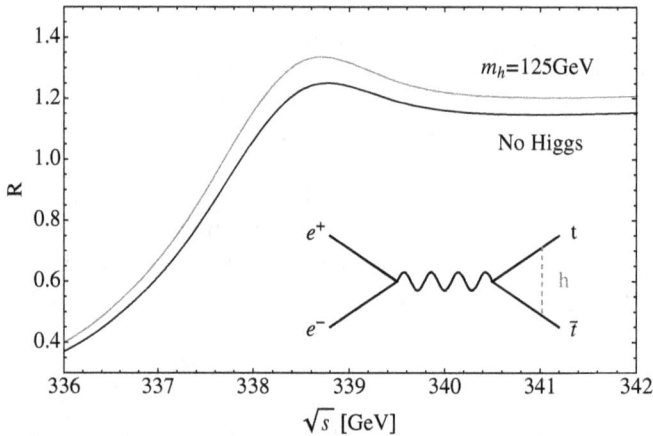

Figure 3.19. Comparison of the cross sections for the $t\bar{t}$ pair production with and without a diagram with the Higgs boson exchange in the loop. The value $m_t = 170 GeV$ was assumed. Source: G Moortgat-Pick *et al* 2015 *Eur. Phys. J.* **C75** 371, figure 105.

3.2.3 Top couplings

Another potentially important measurement that can be performed at a linear e^+e^- collider with much higher accuracy than on the LHC, is the measurement of the couplings of the top quark with the Z boson and the photon, $t\bar{t}Z$ and $t\bar{t}\gamma$. In contrast to hadronic colliders, the vertices $t\bar{t}Z$ and $t\bar{t}\gamma$ are included directly in the leading tree diagrams of the process $e^+e^- \rightarrow t\bar{t}$. The main task of the analysis is the separation of vector and axial-vector contributions in the $t\bar{t}Z$ and $t\bar{t}\gamma$ vertices. Experimentally, this can be done by collecting data with different beam polarization conditions, by measuring the asymmetries in the direction of the top quark production A_{FB}, and polarization of the top quarks. The top quark polarization can be measured using the angular distributions of the decay products, since the top quark decays before its hadronization and, therefore, the information on the spin direction is not lost.

From a theoretical point of view, the measurements of the couplings $t\bar{t}Z$ and $t\bar{t}\gamma$ are very important because they are sensitive to BSM model contributions. In particular, a significant modification of these couplings is predicted within the framework of the Composite Higgs models. In some versions of this approach the couplings $t\bar{t}Z$, and $t\bar{t}\gamma$ will clearly deviate from the SM values. These effects with significance larger than $3\,\sigma$ can be observed with a data sample 500 fb^{-1} collected at 500 GeV.

Many BSM models predict a Z' boson, which is a heavier analog of the Z boson. In particular, the Z' boson must exist in different versions of the Great Unification Theory (GUT), based on the SO(10) group. Similar to the Z boson, the Z' boson will be a mediator in the $e^+e^- \to t\bar{t}$ process. Therefore, the measurements of the $e^+e^- \to t\bar{t}$ process will be sensitive to this particle, if its mass is not too large.

There are a number of other important measurements of the top quark properties, which can be performed at the e^+e^- colliders. The mass and width of the top quark can be reconstructed using its decay products. The searches for rare top quark decays are possible for channels with large enough branching fractions and low backgrounds. In particular, the decays $t \to Ws$ and $t \to Hc$ can be observed. The measurement of the mass difference between the top and anti-top quarks is an interesting topic, providing a direct test of the CPT conservation law.

Finally, it should be noted that the study of the top quark at e^+e^- colliders should be performed in three collision energy regions. First, a data sample of ~100 fb^{-1} near the $t\bar{t}$ mass threshold should be collected. Then a few hundred fb^{-1} should be taken in the peak of the $t\bar{t}$ production cross section at ~380 GeV. It would also be very useful to collect a data sample of a few hundred fb^{-1} at an energy (500–550) GeV, where the channels with three heavy particles in the final state can be observed.

3.3 Additional research topics at ILC

At 250 GeV various additional measurements can be performed at the ILC. First of all, a huge amount of the Z and W bosons will be produced, which allows us to measure their behaviours with high precision. Specific processes with the b quarks and τ leptons attract some interest too. Potentially, searches for NP phenomena are possible. Unfortunately, the chances to find a heavy NP particle at the ILC at 250 GeV are very low. However, the sensitivity to indirect manifestations of NP effects is quite high. New theoretical ideas will likely be proposed during the next decade; these could be experimentally tested at the ILC. Below, the three important measurements are discussed, those that can be performed at the ILC with 250 GeV, complementary to the Higgs boson research program.

3.3.1 Two fermion production channels

If a heavy NP particle has very small couplings with the SM particles, a high luminosity e^+e^- collider, such as the ILC, could be the best place to search for an indirect effect induced by the particle. A detailed study of the processes $e^+e^- \to f\bar{f}$, where the fermions f are all possible quarks and leptons, can be sensitive to NP effects. The general approach to experimental studies of the $e^+e^- \to f\bar{f}$ processes is very similar to that for the $e^+e^- \to t\bar{t}$ process studies discussed in the previous section. However, in contrast to the $t\bar{t}$ production case, the studies of the $e^+e^- \to f\bar{f}$ processes are possible even at the ILC with 250 GeV.

The dominant SM diagrams describing the processes $e^+e^- \to f\bar{f}$ include the vertices $f\bar{f}Z$ and $f\bar{f}\gamma$. Therefore, the corresponding couplings can be measured, and the vector and axial-vector contributions to these couplings can be separated, similar to the $e^+e^- \to t\bar{t}$ process. Using this method, the deviations of the couplings from the SM predictions can be searched for.

Such deviations can be manifested in models with new gauge bosons, composite fermions, and additional spatial dimensions. Again, the most interesting effect, which can be clearly tested in these measurements, is the inclusion of the diagram with the Z' boson exchange in addition to the diagrams with the γ and Z boson exchange.

The search for the Z' boson was performed at the LHC and an upper limit on its mass of about ~3 TeV was set. At the ILC, the method of separating the contribution from the $f\bar{f}Z'$ vertex from the dominant contribution of the $f\bar{f}Z$ and $f\bar{f}\gamma$ vertices is similar to that described in the previous section on the top quark coupling studies. The cross sections of the processes have to be measured with different beam polarizations. The angular distributions of the final state particles have to be checked. To find the contribution of the Z' boson, it is important to measure the pair production of quarks and leptons with all possible flavors. As a result of such a study at the ILC, it is possible to determine contribution or to obtain an upper limit on the existence of the Z' boson with a mass of up to 5 TeV, and up to 10 TeV in some BSM models.

3.3.2 The search for NP particles

The next method of searching for NP at 250 GeV is the study of the spectrum of radiative photons. As a result of the ISR, energetic photons can be produced, and their energy distribution is sensitive to NP. As an example of such study, a search for weak interaction massive particles (WIMP), motivated by the phenomenon of dark matter, can be performed. Using this method, the contribution of the process $e^+e^- \to \chi\chi\gamma$ can be evaluated, where a WIMP particle χ with a mass up to 100 GeV is not detected. In this method, the missing mass to the radiative photon is measured. A bump in the monotonous distribution of the radiative photons will indicate the contribution of the process with the WIMP particle. Monochromatic energy ISR photons can be searched for too because they can appear in specific models of NP.

Similar to the previous method, not very heavy NP particles X with a mass of a few tens of GeV can be searched for in the process $e^+e^- \to ZX$. For this, the missing mass distribution to the Z boson has to be studied. This method was used in the LEP experiments, however the sensitivity at ILC will be much higher. The proposed analysis will limit the number of BSM models even in the case of a negative result.

These methods of searching for NP are possible even at 250 GeV. However, these methods will work better at higher energies, in particular at 500 GeV. Additional methods of searching for new particles and phenomena at e^+e^- linear colliders were discussed in the literature, however, they usually require energy above 250 GeV.

3.3.3 Pair production of W and Z bosons

It is important to measure the parameters of the electroweak theory with high accuracy. Such studies were performed at the LEP experiments, however, there are several reasons why these measurements must be repeated at the ILC. First, the polarized beams at the ILC provide new opportunities, in particular, for the background suppression. Secondly, the full integrated luminosity at the ILC is expected be about 1000 times greater than in all four LEP experiments. Thirdly, the ILD and SiD detectors at the ILC will have significantly higher accuracy, for example, momentum resolution in the tracker will be better by 1–2 orders of magnitude. Fourthly, there is some inconsistency in the SM parameter values, which requires the new measurements with higher accuracy.

At the ILC with 250 GeV, the processes $e^+e^- \to W^+W^-$ and $e^+e^- \to ZZ$ will have large cross sections and, respectively, a large number of events. Measurements of these processes will allow us to improve the accuracy of the mass, width, decay branching fractions, and couplings measurements of the W and Z bosons. It will also be possible to study the structure of the couplings γWW and ZWW. In many BSM models, some deviations of these couplings from the SM values are possible. In the EFT formalism, NP effects can be described by the tensor interactions introduced in a form similar to equation (3.13).

In addition to the two-particle channels, a single W boson can be produced in the $e^+e^- \to W^{\pm}e^{\mp}\nu_e$ process, however, its cross section is small. At the next stage of the ILC with a higher collision energy, it is possible to study the channels with three vector bosons $e^+e^- \to ZW^+W^-$ and $e^+e^- \to ZZZ$. In addition to a theoretical interest, it is important to measure the cross sections with gauge bosons production to estimate the backgrounds for other studied processes.

One of the potentially important measurements at the ILC is the precise measurement of the mass of the W boson. At 250 GeV, the W boson mass can be reconstructed using the decay products and taking into account the kinematics of the process. Systematic uncertainty will dominate, since a large statistics of the W bosons will allow us to get a low statistical uncertainty. Using this method at the ILC, it is possible to measure the W boson mass with an accuracy of a few MeV. It is very important to obtain such a high accuracy because this uncertainty is of the same order of magnitude as the loop corrections induced by NP particles, which appear in different BSM models.

Chapter 4

Conclusion

In this book the linear e^+e^- collider ILC project at 250 GeV and higher is discussed. The ILC project is well developed from the point of view of collider and detector technologies, as well as the physical research program. The project is very actual, in the case of construction the ILC will determine the direction of development of fundamental particle physics theory for the next decades. The cost of the project is about 5.5 billion dollars, however, the spending will be distributed over 10 years of construction and among many participating countries. In addition to potential progress in fundamental physics, valuable technology development is expected, which can lead to practical applications.

At the end of 2018 a decision on the construction of the ILD collider in Japan is expected. However, even in the case of a negative decision, the linear e^+e^- collider project is so actual that it will be realized sooner or later in that or another country. Let's hope that the construction of the ILC collider in Japan will begin in the near future.

4.1 Additional resources

- A detailed description of the technical aspects of the ILC collider, the ILD and SiD detectors and the proposed program of physics research can be found in four volumes of the project TDR. It can be found in:
  - Behnke T *et al* 2013 *The International Linear Collider Technical Design Report, Volume 1: Executive summary* (arXiv:1306.6327).
  - Baer H *et al* 2013 *The International Linear Collider Technical Design Report, Volume 2: Physics* (arXiv:1306.6352).
  - Adolphsen C *et al* 2013 *The International Linear Collider Technical Design Report, Volume 3: Accelerator* (arXiv:1306.6328).
  - Behnke T *et al* 2013 *The International Linear Collider Technical Design Report, Volume 4: Detectors* (arXiv:1306.6329).

- I recommend to students, who are interested in the Higgs and top physics, for a pedagogical reason, to read the corresponding chapters in PDG:
  - Higgs physics: http://pdg.lbl.gov/2018/reviews/rpp2018-rev-higgs-boson.pdf
  - Top quark physics: http://pdg.lbl.gov/2018/reviews/rpp2018-rev-top-quark.pdf

- A comprehensive program of physics research at ILC is discussed with some emphasis on the Higgs boson studies in the following reviews:
  - Weiglein G *et al* 2006 LHC/LC Study Group: Physics interplay of the LHC and the ILC *Phys. Rep.* **426** 47.
  - Brau J E *et al* 2012 The Physics Case for an e^+e^- Linear Collider (arXiv:1210.0202).
  - Asner D M *et al* 2013 ILC Higgs White paper (arXiv:1310.0763).
  - Moortgat-Pick G *et al* 2015 Physics at the e^+e^- linear collider *Eur. Phys. J.* **C75** 371.
  - Fujii K *et al* 2015 Physics Case for the International Linear Collider (arXiv:1506.05992)
  - Barklow T *et al* 2018 Improved formalism for precision Higgs coupling fits *Phys. Rev.* D **97** 053003.
  - Fujii K *et al* 2017 Physics Case for the 250 GeV Stage of the International Linear Collider (arXiv:1710.07621).

- Some specific details of the top physics at a linear e^+e^- collider are given in:
  - Atwood D *et al* 2001 *CP* violation in top physics *Phys. Rep.* **347** 1–222.
  - Seidel K, Simon F, Tesar M and Poss S 2013 Top quark mass measurements at and above threshold at CLIC *Eur. Phys. J.* **C73** 2530.
  - Vos M *et al* 2016 *Top physics at high-energy lepton colliders* (arXiv:1604.08122)
  - Vos M 2017 Top physics beyond the LHC (arXiv:1701.06537).

www.ingramcontent.com/pod-product-compliance
Lightning Source LLC
Chambersburg PA
CBHW082113210326
41599CB00033B/6683